PREMISES CABLING

Donald J. Sterling, Jr.

Delmar Publisher's Online Services
To access Delmar on the World Wide Web, point your browser to:
http://www.delmar.com/delmar.html
To access through Gopher: gopher://gopher.delmar.com
(Delmar Online is part of "thomson.com", an Internet site with information on
more than 30 publishers of the International Thomson Publishing organization.)
For information on our products and services:
email: info@delmar.com
or call 800-347-7707

Delmar Publishers

I(T)P An International Thomson Publishing Company

Albany • Bonn • Boston • Cincinnati • Detroit • London • Madrid
Melbourne • Mexico City • New York • Pacific Grove • Paris • San Francisco
Singapore • Tokyo • Toronto • Washington

NOTICE TO THE READER

Publisher does not warrant or guarantee any of the products described herein or perform any independent analysis in connection with any of the product information contained herein. Publisher does not assume, and expressly disclaims, any obligation to obtain and include information other than that provided to it by the manufacturer.

The reader is expressly warned to consider and adopt all safety precautions that might be indicated by the activities described herein and to avoid all potential hazards. By following the instructions contained herein, the reader willingly assumes all risks in connection with such instructions.

The publisher makes no representations or warranties of any kind, including but not limited to, the warranties of fitness for particular purpose or merchantability, nor are any such representations implied with respect to the material set forth herein, and the publisher takes no responsibility with respect to such material. The publisher shall not be liable for any special, consequential, or exemplary damages resulting, in whole or in part, from the readers' use of, or reliance upon, this material.

Delmar Staff:
Executive Director: Dale Bennie
New Product Development Manager: Mark Huth
Editor: Jack Erjavec
Production Manager: Dianne Jensis
Marketing Manager: Kathryn Little

Copyright © 1996
By Delmar Publishers
a division of International Thomson Publishing Company

The ITP logo is a trademark under license.

Printed in the United States of America

For more information, contact:

Delmar Publishers
3 Columbia Circle
Albany, New York 12212-5015

International Thomson Publishing Europe
Berkshire House 168-173
High Holborn
London WC1V 7AA
England

Thomas Nelson Australia
102 Dodds Street
South Melbourne, 3205
Victoria, Australia

Nelson Canada
1120 Birchmount Road
Scarborough, Ontario
Canada M1K 5G4

International Thomson Editors
Campos Eliseos 385, Piso 7
Col Polanco
11560 Mexico D F Mexico

International Thompson Publishing GmbH
Königswinterer Strasse 418
53227 Bonn
Germany

International Thomson Publishing Asia
221 Henderson Road
#05 - 10 Henderson Building
Singapore 0315

International Thomson Publishing Japan
Hirakawacho Kyowa Building, 3F
2-2-1 Hirakawacho
Chiyoda-ku, Tokyo 102
Japan

1 2 3 4 5 6 7 8 9 10 XXX 01 00 99 98 97 96 95

Library of Congress Cataloging-in-Publication Data

Sterling, Donald J., 1951-
    Premises Cabling / Donald J. Sterling, Jr.
       p. cm.
    Includes index.
    ISBN 0-8273-7244-2
    1. Telecommunications wiring. 2. Office buildings--Electric
equipment. 3. Commercial buildings--Electric equipment. I. Title.
TK5103.12.S74 1995
621.39'81--dc20
                                95-31998
                                CIP

Contents

Forward

I've been searching for a versatile book to use as a reference text for designers and installers of premises structured cabling systems. The industry has provided standards such as TIA/EIA-568A and ISO/EIC 11801 that detail specific requirements, but these don't provide the necessary framework for the uninitiated.

This book begins (in Chapter 1) with introductory information that illustrates the "state-of-the-art" today through explanations of contemporary trends and global standards activity.

Immediately following the introduction, you delve into specific issues related to copper and fiber cables and components (Chapters 2, 3, and 4). These chapters not only define essential terminology, but include narrative that fleshes out the reader's understanding of the cabling systems subject. An example is the explanation of "Megahertz versus Megabits." This clarification helps the user to see why a Category 5 link certified to 100 MHz can actually transport data at 155 Mbps. You also get an industry observer's view of the "Fiber versus Copper" question.

Chapters 7, 8, and 9 easily follow with detail on installation, certification, and administration of the installed cable plant. Chapter 7 addresses the everyday practices that must become second nature to the cable system installer. Chapter 8 shows you how to demonstrate that the installed cable plant will perform according to plan and, if it doesn't, what to do to correct the problem. Chapter 9 emphasizes the often taken for granted practice of cable system administration through labeling and documentation.

Although these chapters are fundamental to the book's purpose, Chapters 5 and 6 provide enlightening answers to questions asked daily in the industry by those who are active participants. Chapter 5 deals with the networks that run on the cabling systems and explains how the "bits" fit into the picture. Chapter 6 takes the cables and connectors discussed in Chapters 2 through 4 and puts them into a building setting, where you begin to visualize the cable plant as a significant asset in the overall building structure.

Whether you are interested in background information related to premises structured cabling systems or just in understanding specific cabling system issues, this book provides the first real reference book for those wanting to actively participate in the high-performance cabling systems industry.

Gregory J. Weldon
AMP Incorporated

Preface

Premises Cabling is simple in purpose: to give a general survey of standards-based generic cabling for buildings. To a large degree, the book grew from my own needs to gather together much of the information presented.

Because of the trend toward standards-based cabling, the book gives TIA/EIA-568A a great degree of emphasis and ISO/IEC 11801 a lesser degree. One challenge of the book was to keep up with the evolution of these documents as they went through numerous revisions on their way to approval.

The first part of the book deals at the component level with cables and connecting hardware. Then to change pace, Chapter 5 deals mainly with networking. Because building cable evolved mainly to meet the needs of networks, I've tried to give some perspective on how networks fit into building cabling schemes – or, perhaps, how building cabling schemes fit into networks.

The remainder of the book considers cabling systems from a system-oriented viewpoint, with some guidelines on how to install it, test it, and administer it.

Of the many people who helped with advice, counsel, and perspective, a few deserve special mention. Bret Matz, Michael Cohen, and Rick Downs of Connectware helped confuse me in the first place. This book, in part, is an attempt to get it all straight. Les Baxter of AT&T Bell Laboratories and Greg Weldon of AMP Incorporated both provided valuable information and reviewed the manuscript, offering sage advice and comments that I sometimes ignored at my peril. The standard disclaimer applies: despite the best efforts of others, the mistakes remain my own.

Lastly, my wife Lynne and daughter Megan put up with the hectic schedule of bookmaking. Patience is indeed a virtue. Megan, though, thought bookmaking might be cool if her name was in it. So this book's for her:

Megan Owen Sterling

CHAPTER 1

Introduction

Not too many years ago, wiring a building was a simple affair. Electricians wired the building for power. Telephone craftspeople wired the building for telephone service. If there were any large computers, they were usually wired together by the computer company that supplied the hardware. Lines of responsibility were clearly delineated, with little confusion or overlap.

The personal computer changed this orderly approach. Invented in the 1970s and widely adapted by business in the 1980s, the personal computer quickly became a fixture on the modern office desk. Fact is, after the telephone, the PC is probably the most common piece of equipment in the office. People share copiers, printers, fax machines, and even coffeemakers. But the PC is an individual thing; like the telephone, it is installed on every desktop. What is shared is data over the network, not the computer hardware itself.

The PC by itself did not change the wiring of buildings: the networks, too, played a role. Indeed, networks predate the PC. Xerox invented Ethernet before the PC existed. IBM released its cabling system before it released either its PC or its Token Ring network. The earliest adopters of networks used workstations (essentially a very powerful and quite expensive desktop computer) and minicomputers.

If computers are neat and powerful tools allowing individuals to become more creative and productive, then linking them together brought the same benefits to the workgroup. People could share information, send electronic messages, and share resources like printers, fax machines, and mass storage.

The PC and network sowed confusion in how a building is wired. Different networks had different cabling requirements. Worse yet, computer networks tied together into larger networks (or large networks were subdivided into smaller networks), along with the need to link different types of networks. Remotely separated networks were linked together over the public telephone system (itself the world's largest electronic network). The huge mainframe and smaller sibling, the minicomputer, were hastily and incorrectly declared dinosaurs on the edge of extinction. The network was hailed as the future of computing.

Here was the crux of the problem in the 1980s: in the 1960s and '70s, a company committed to a certain type of computer – say, an IBM mainframe. IBM then took complete responsibility for the computer installation, including the cabling that connected both peripherals and hundreds of terminals to the mainframe. But with the rise of the PC and networks, no single vendor had responsibility. IBM and a few other companies had the vision to offer a single-source approach to the system, but buyers were wary of being tied to a single vendor. Companies bought PCs from IBM, Compaq, Apple, and a dozen other companies; printers from Hewlett-Packard, Apple, and QMS; network software from Novell and Banyan; network hardware from Synoptics, 3Com, and other startups in this new field; and application software from Microsoft, Lotus, and WordPerfect. (And this namedropping only hints at the number of vendors available!)

Suddenly, there was no single point of responsibility for either the network or the cables that interconnected everything. As networks became standardized, with well-defined rules, cabling did not quite follow suit so readily. In the old days, if a problem occurred with your cabling, you called the electrician, the phone company, or the computer company. With the network, who do you call? "Not my job", said the computer vendors, the network hardware suppliers, and just about everybody else involved. "I don't know about networks", said the architect. In such a situation, everyone began pointing fingers and claiming it was the other guy's equipment causing the problem.

Large companies developed in-house expertise. Those involved with creating and running the network became intimately involved in understanding the cabling. (A sign of the new sophistication: the *wiring* became the *cabling*.) Independent computer/network consultants were also available. Most were quite good and competent, while some presented a dubious gamble at best.

THE PHYSICAL PLANT IS CRITICAL

Most problems in a network can be traced to the so-called physical plant, principally the cables and connectors that tie everything together. By various estimates, anywhere from 50% to 75% of all network problems involve the physical plant. Regardless of whose estimates you use, the importance of reliable cabling systems is obvious. If a failure occurs, first check the cabling system. Disruptions occur from many sources:

- Shoddy workmanship during installation. Incorrectly applied connectors, poor cable routing, cable kinks and excessive bends are among the most common problems. This is a particularly frequent problem with newer high-speed networks.

- Violations of interconnection rules involving allowable cabling distances and types of cables used.

- Use of substandard components in relation to the demands of the application.

- Failure to maintain the plant as circuits are added, moved, or rearranged.

- Human error in installing equipment and maintaining the plant.

- Having a constrained, rigid system whose growth and evolution can only be accommodated by Rube Goldberg renovations.

- Users who attempt to make the cabling outperform its capabilities or who incorrectly use the wrong cable. For example, using a low-speed telephone cable to make a high-speed network connection sometimes works, but usually doesn't.

- Electrical noise from surrounding equipment, including power cables, fluorescent lights, motors, and so forth.

- Rodents chewing through cable. Buildings have mice, and they will nibble on cable.

STANDARDS ARE US

A related trend was the powerful move toward standards-based open systems. Companies no longer wanted to be tied to single-company, proprietary schemes. They wanted choice, and safety in the knowledge that the equipment they bought from one company would work with the equipment purchased from another company. The PC architecture became a *de facto* standard. Networks, too, were standardized by bodies like the Institute of Electronic and Electrical Engineers (IEEE), American National Standards Institute (ANSI), International Electrotechnical Committee (IEC), and International Standards Organization (ISO).

So why not standardize cabling systems? AT&T and IBM both offered structured cabling systems. These offered two significant benefits: flexibility and coherency. By coherency, we mean that the system was, in fact, a system tied together in an orderly and rational fashion.

Coherency, in turn, provided flexibility for moves, adds, and changes. One thing in building cabling is certain: it isn't static. People change offices, workgroups change personnel, and companies grow (or contract) as they succeed (or don't succeed). A structured system or open system allows moves, adds, and changes in wiring to be done easily. Rather than ripping out the wiring (a practice more common than you might think) and installing new cabling, simple changes at different points in the system could be easily made by simply rearranging components or plugging in new ones. Equally important is the idea of "futureproofing" a building against premature obsolescence of the cable plant. The cabling should meet tomorrow's needs as well as today's.

The IBM cabling system had limited success, but it largely failed because it did not adequately meet the evolving needs of users. It still seemed too focused on IBM equipment and systems: its greatest popularity is with the IBM-backed Token Ring network. And the shielded twisted-pair cable favored is undeniably a good performer. It is, however, being edged out by lower cost alternatives that still meet the performance requirements of different applications. In other areas, IBM guessed wrong; the type of optical fiber they specified is different from the type adopted by the rest of the industry. To its credit, IBM has moved away from its own cabling system and is now adopting open cabling systems.

AT&T's SYSTIMAX structured cabling system was built on twisted-pair and fiber-optic cables and components such as cross connects that grew out of its telecommunications experience. Twisted-pair and fiber-optic cable are the preferred cabling methods today, so the AT&T system is compatible with premises cabling standards.

As networks and cabling became more open, other vendors like connector-giant AMP developed strategies to allow the widest flexibility in wiring. AMP's approach in its NET-CONNECT Open Cabling System is a modular one that allows modules to be interchanged to meet specific wiring needs. While the AT&T approach is built on preferred types of cables, the AMP system is designed to accommodate any type of cabling.

By the beginning of the '90s, the industry was ready for standards defining how buildings should be cabled for both telephone and data. Both TIA/EIA-568, in the United States, and ISO/IEC 11801, internationally, address cabling of commercial buildings. The importance of these standards is that they are *performance* driven rather than *application* driven. They don't recommend cabling specifically for Ethernet or Token Ring. Rather the guidelines recommend methods of installing cable for, say, 20-MHz operation. This cabling, if properly installed, will permit either (or both) Ethernet or Token Ring to run on it. The hot area in networking today is high-speed networks operating at 100 to 155 Mbps. By installing a 100-MHz cabling system, you don't have to know what network you are going to use. The cabling should accommodate FDDI, ATM, Fast Ethernet, 100VG-AnyLAN, or even a network scheme not yet created.

Newer networks are more open in their cable requirements; that is, they run on common twisted-pair or fiber-optic cable, rather than special cables specific only to a given network. Both the TIA/EIA and ISO/IEC standards recognize the need to support "legacy" LANs, those with cabling needs that predate and differ from those recommended by the standards. For example, the earliest versions of Ethernet used coaxial cable. While some users with heavy investments in coaxial Ethernet may prefer to continue with coaxial cable, few new installations will choose to do so. Twisted-pair cable – cheaper, smaller, and easier to work with – has largely replaced coaxial cable. A structured cabling system should be able to meet both the application-independent requirements of TIA/EIA-568A and ISO/IEC 11801 and those of legacy applications.

Because TIA/EIA-568A and ISO/IEC 11801 will become dominant standards for cabling buildings, we will spend a fair portion of this book looking at their requirements. The two standards are very similar, but differ in some important points. It is possible to certify performance to one standard and fail to meet the requirements of the other. TIA/EIA-568A is a subset of ISO/IEC 11801. With a few minor exceptions, following 568 also means meeting the requirements of 11801. The converse is not true. The 11801 standard permits things that the 568A standard does not.

One important thing to remember about 568A and 11801 is that they represent the consensus of experts from both the telecommunications and network industries. Many of the same people who worked to create the cabling standards for commercial buildings also serve on committees defining network and telecommunications standards. Thus the standards represent an important effort to accommodate the real needs of users. Building cabling standards, on one hand, attempt to create a flexible structure to satisfy the needs of emerging applications. These same emerging applications are defined with the cabling standards in mind. For example, a significant design feature of 100VG-AnyLAN was to allow high-speed network communications over modest cabling already installed in the building. The network protects the investment many companies have in their present wiring and offers increased performance without the need to recable. Users don't want a network that requires special cabling, or one that isn't open and flexible enough to meet growing and evolving needs.

But before we begin looking at the elements of building cable in detail, this chapter will offer a quick overview of building cabling and of networks. Regardless of the application independence of building cabling standards, a practical approach must consider the needs of the network. Because we will look at networks in some detail later and will refer to them throughout, we'll end this chapter with a brief overview of some popular networks.

THE BANDWIDTH CRUNCH

A recurring consideration with networks is to provide enough bandwidth to accommodate all users. As the needs of more and more users became more sophisticated, low-speed networks became a bottleneck in passing information from station to station. If users are only trading files and messages of modest size, a low-speed network suffices. But applications quickly grew until they started consuming a significant portion of the bandwidth. For example, the word processing file of the text for this chapter is only a modest 28,000 bytes. A magazine ad with a large color photograph can run to several million bytes. Computer-aided design files can also consume millions of bytes. Multimedia, and especially video, also produce large files running to millions of bytes.

Applications like video and multimedia also place demands on networks. Not only do they require a high bandwidth, they also are real-time applications. They cannot be delayed by other files being transferred or by bottlenecks in the network, at least not if you want something more than jerky, flaky images.

The quest for bandwidth leads naturally to higher speed networks. Early networks operated at 16 Mbps or less, while emerging networks run at 100 and 155 Mbps. Other techniques, which we will discuss in Chapter 5, also make more efficient use of bandwidth.

The point is this: as networks operate faster, the demands they place on the cabling system also increase. A high-speed network is a trickier animal to tame than a low-speed network. As a result, the cable plant and its proper installation become increasingly critical to the reliable operation of the network.

BUILDING CABLING

The cabling in a building can be divided into two sections (Figure 1-1): the backbone cabling running between floors and between buildings and the horizontal cabling running to different offices on a floor. Cables are usually consolidated at different points and then distributed through the buildings. These consolidation points are affectionately called *wiring closets*, although in some large installations they make quite a spacious closet.

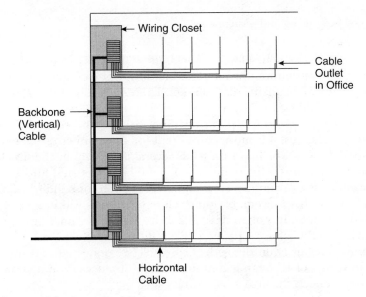

Figure 1-1. Backbone and horizontal cabling

There are three levels of distribution: the main cross connect, intermediate cross connect, and horizontal cross connect. Depending on the size of the building, not all three levels are necessary. Figure 1-2 shows the hierarchy of wiring closets.

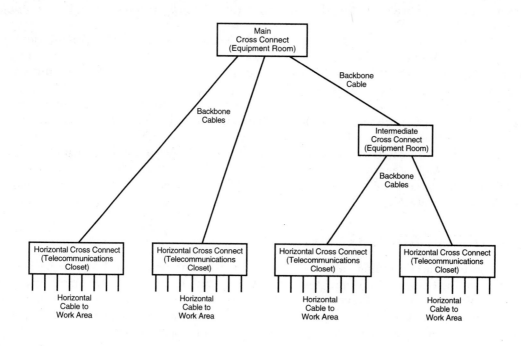

Figure 1-2. Hierarchy of cabling and wiring closets

The *main cross connect* is the first level of backbone cable. It is the area where outside cables enter the building for distribution through the building. For example, the outside telephone lines will be terminated in the main cross connect. There is usually one main cross connect per building or group of buildings.

The *intermediate cross connect* is a point dividing the first and second levels of backbone cabling. In large installations, it is more convenient for the main cross connect to feed to an intermediate cross connect, which in turn feeds several telecommunications closets.

The *horizontal cross connect* is the point at which the backbone and horizontal cables meet.

Backbone cables also can connect cross connects in different buildings. Some companies, schools, and universities have several buildings close together – within several hundred yards – in what is called a campus. It is often desirable to have all these buildings essentially wired as a single entity. When buildings are separated by significant distances – as far apart as corporate offices in different cities or countries – they can still be connected, but usually though the public telephone system or through leased lines. Such distances do not allow direct connection of main cross connects through a backbone cable.

We should mention that TIA/EIA and ISO/IEC standards are recommendations; they do not have the force of law. You can cable your building any way that works. You can use cabling rules from a specific vendor, you can design your own system, or you can run your cables and cross your fingers. But by following widely recognized standards, you simplify planning and ensure a cabling plant that will perform to specification.

NETWORKS

A local area network is simply a means of interconnecting computers and related equipment like printers. That's simple in concept; a lot of ingenuity went into making networks practical. Over the years a number of different networks have evolved. From the perspective of premises cabling, the two most important things about networks are the types of cables they require and the speed at which they operate. Indeed, the trend in networking has been to simplify the cabling while increasing the operating speed. Figure 1-3 gives a short summary of the networks, their operating speed, the types of cables they require, and the approximate year they were introduced. Notice that all network types use twisted-pair cable and optical fibers. FDDI and ATM are notable in that they are networks specifically designed to use optical fibers; other networks have been adapted to allow use of fiber.

Network	Cable Types	Operating Speed	Standard	Year
Ethernet	Unshielded twisted pairs Coaxial Optical fiber	10 Mbps	IEEE 802.3	1980
Token Ring	Shielded twisted pairs Unshielded twisted pairs Optical fiber	4 Mbps/ 16 Mbps	IEEE 802.5	1985
FDDI	Optical fiber Unshielded twisted pairs Shielded twisted pairs	100 Mbps	ANSI X3T9.5	1990
Fast Ethernet	Unshielded twisted pairs Shielded twisted pairs Optical fiber	100 Mbps	IEEE 802.3	1995
100 VG–AnyLAN	Unshielded twisted pairs	100 Mbps	IEEE 802.12	1995
ATM	Optical fiber Twisted pairs	25, 51, 100 and 155 Mbps	ATM Forum	1995

Figure 1-3. Popular types of networks

(The dates indicate the year a substantially complete specification was published for the network. Products for the type of networks were marketed and networks built before that date. However, these did not necessarily meet the final published specifications. Additional details may be published later. For example, ATM devices were built in 1993 and 1994, but the final specification for local-area ATM networks was not approved until 1995. The earliest adopters faced the possibility of ending up with equipment that did not meet specifications. Remember, too, that networks continue to evolve.)

Ethernet was the earliest network to gain widespread popularity; today it is available in more flavors than any other. Operating at 10 Mbps, Ethernet is the most widely used network today.

Token Ring was popularized by IBM. Originally operating at 4 Mbps and later at 16 Mbps, Token Ring networks follow Ethernet in popularity, principally because of the huge installed base of IBM equipment.

FDDI (Fiber Distributed Data Interface) was the first network specifically designed for fiber optics. It was also the first network to operate at 100 Mbps. The network can also use copper cables in certain segments. FDDI never became as popular as early proponents hoped and may be eclipsed by newer competing approaches.

Fast Ethernet is a significant upgrade to bring 100-Mbps speed to Ethernet networks. We separate it from other Ethernet flavors because it does not operate at 10 Mbps. Fast Ethernet has two main variations, termed 100BASE-X and 100BASE-T4.

100VG-AnyLAN is a network promoted by Hewlett-Packard, AT&T, IBM, and several other vendors. Operating at 100 Mbps, its distinguishing feature is its ability to operate on a lower grade of unshielded twisted-pair cable than other 100-Mbps LANs. What's more, it offers a migration path for both Token Ring and Ethernet.

ATM (Asynchronous Transfer Mode) is perhaps the network that will have the strongest long-term significance. ATM operates at speeds up to 2.4 Gbps (although in the local area network speeds range from 25 Mbps to 155 Mbps). Equally important, it is compatible with high-speed fiber-optic telephone networks, making its integration in long-distance networking and telecommunications more practical and straightforward. As this book is being written, ATM has gained much more attention than application. Final standards have not been worked out, but the shape of ATM is clear enough.

BUILDING CABLING: THE SYSTEM

There are several considerations involved in planning and implementing a premises cabling system:

1. Choose the level of performance required. This will determine the type of cable required.
2. Decide what installation rules to follow: TIA/EIA-568A, for example.
3. Decide whether any legacy applications will need to be supported. This will determine if any additional cabling considerations apply.
4. Install the cable, being careful to follow applicable applications guidelines and practices.
5. Certify the cabling plant. This means testing the installation to ensure that the performance meets requirements.
6. Document the cabling. When moves, adds, changes, or repairs are required, you'll need to know what's what and what's where.
7. Use the system (enjoyably).

CHAPTER 2

Copper Cables

Since cables form the heart of the premises cabling system, we will look closely at the types and properties of cables in detail. The great division among cables is between copper cables carrying electrical signals and optical-fiber cables carrying light signals. This chapter covers copper cables; the next chapter looks at optical cables.

Many cables have been developed or adapted for network and premises wiring applications. The two main categories of cable are *coaxial cable* and *twisted-pair* cable. A cable has a simple purpose: to carry a signal a given distance at a given data rate. In doing so, the cable must protect the signal from distortion due to noise. Sufficient noise can turn a meaningful signal into gibberish, cause circuits to trigger falsely, ruin an important file, or crash a computer. Because cables are designed to minimize noise and ensure the quality of the transmitted signal, we'll begin with a look at noise.

NOISE

When the first personal computers were introduced in the late 1970s, they radiated enough energy to interfere with television several hundred feet away – not just in the house where it was used but in neighboring houses as well. This interference is known as electromagnetic interference (EMI). EMI is a form of environmental pollution with consequences ranging from the merely irksome to the deadly serious. EMI making television reception snowy is one thing; its interfering with a heart pacemaker is quite another story.

A copper wire can be a main source of EMI. Any wire can act as an antenna, radiating energy into the air where it can be received by another antenna. A simple demonstration of EMI can be done by placing an AM radio near a computer. You can tune into the computer at some point on the dial. Try running different programs to hear the differences each makes. What you are hearing is EMI.

Just as a wire can be a transmitting antenna, it can be a receiving antenna, picking up noise out of the air. Obviously, this received noise can interfere with the signal being carried.

CROSSTALK

A special type of noise is crosstalk. Crosstalk is energy coupled from one conductor to another in the same cable or between cables. Crosstalk was once fairly prevalent in telephone systems, where you could faintly hear another conversation.

When current flows though a wire, it creates an electromagnetic field around the wire. This field can induce a current in an adjacent wire within the field. This coupling is crosstalk.

Figure 2-1 shows the idea of crosstalk. A signal is transmitted down a wire (called the active or driven line). A certain portion of the energy couples on a second wire (called the quiet line). This energy on the quiet line is the crosstalk. Crosstalk is measured either in percentages or decibels. In percentages, the magnitude of the signal on the quiet line is given as a percentage of the signal on the driven line. Thus the crosstalk might be 3% or 10%. If the driven line signal was 5 volts, a 10% crosstalk means that 0.5 volt appeared on the quiet line. In communication cabling, crosstalk is more often expressed in decibels, a term we'll discuss in detail shortly.

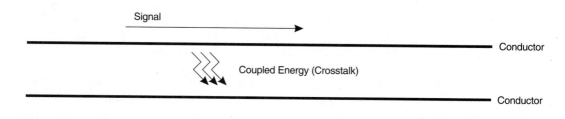

Figure 2-1. Crosstalk

Crosstalk can be measured at the near end or far end of the line. In near-end crosstalk (NEXT), the energy on the quiet line is measured at the same end as the source of the signal. In far-end crosstalk, the energy is measured at the opposite end. In premises cabling, NEXT is the type of crosstalk of interest.

Why NEXT? Consider a signal on the driven line. As it travels down the wire, it becomes attenuated, losing energy. At the opposite (far) end, it is weakest. So the greatest amount of energy is available for coupling at the near end. Similarly, crosstalk on the quiet line is usually strongest at the near end for two reasons. First, it can receive the greatest energy here since energy on the active line has not yet become attenuated and, second, it will not be attenuated as it travels to the far end.

NEXT has an additional practical consequence. In a typical network, one pair of wires is used to transmit, while another pair is used to receive, as shown in Figure 2-2. The signal on the transmit line is strongest at the transmission end. A signal on the receive line is also the weakest at this point. There is a greater likelihood of the coupled noise interfering with this weakened signal.

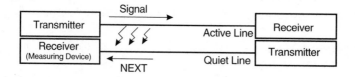

Figure 2-2. NEXT

The formula for calculating NEXT in a cable is

$$\text{NEXT (dB)} = 10 \log \left(\frac{P_q}{P_a}\right)$$

where P_q is the noise power measured on the quiet line and P_a is the driving power on the active line.

There are two basic generalizations that we can make about noise and crosstalk:

Noise and crosstalk become more severe and harder to deal with at higher frequencies than at lower frequencies.

Noise and crosstalk exist in every copper-cable system. The trick is to minimize them to such an extent that they can be ignored.

THE DECIBEL

NEXT loss is measured in decibels. The decibel is used to compare two powers, currents, or voltages. In some cases, it compares the power into the circuit to the power coming out the other end. In the case of NEXT, it compares the power on one line to the power appearing on another line.

The basic equations for the decibel are:

$$dB = 20 \log_{10} (V_1/V_2)$$
$$dB = 20 \log_{10} (I_1/I_2)$$
$$dB = 10 \log_{10} (P_1/P_2)$$

where V is voltage, I is current, and P is power. The reason the decibel is a logarithmic ratio is to allow large variations to be expressed in small units. Each increase of 10 dB for power or 20 dB for voltage or current represents an order of magnitude (X10) change in the ratio.

What does the decibel mean to NEXT? Suppose we have 5 watts of power on the driven cable. A NEXT loss of 10 dB means that 0.5 volt appears on the quiet cable. The 10 dB means that the crosstalk voltage is 90% below the driving voltage. If the NEXT loss is 20 dB, the crosstalk is 99% below the driving voltage. Therefore, the higher the number, the greater the ratio and the lower the energy coupled onto the quiet line as crosstalk.

In dealing with decibels, it's important to keep in mind whether you're looking for a high number or a low number. With NEXT, a higher number is better. It indicates very little energy is coupled onto the quiet line. With the attenuation of a cable, a lower number is better to indicate very little signal power has been lost. A high NEXT number indicates better performance.

Figure 2-3 is a table showing common cable characteristics that are measured in decibels and whether a higher number or lower number indicates better performance.

Characteristic	Better Performance in dB
Crosstalk (NEXT)	Higher
Structural Return Loss	Higher
Cable Attenuation	Lower
Connector Insertion Loss	Lower

Figure 2-3. Putting the decibel in perspective

Another decibel unit you will often come across is the dBm, which means *decibels in reference to a milliwatt*. This simply means that 1 mW is used as a constant in the ratio:

$$dBm - 10 \log_{10} (p / 1 \text{ mW})$$

The dBm unit is widely used in fiber optics to simplify calculations of power in the system. A -10 dBm figure equates to 100 µW of optical power. Furthermore, dBm is preferred over microwatts because it allows losses (in decibels) to be directly added and subtracted. If your cable has a loss of 2 dB, and the connectors account for another 1 dB, then the result is -13 dBm – or 50 µW.

IMPEDANCE

There is one other characteristic that can affect noise and signal transmission quality: impedance. Every cable has a property known as *characteristic impedance*. Characteristic impedance is determined by the geometry of the conductors and the dielectric constant of the materials separating them. For the coaxial cables we will be discussing next, determining characteristic impedance is easy since the cables have a uniform and regular geometry. In other cases, such as twisted-pair cables discussed later in this chapter, characteristic impedance is harder to determine because the geometry is more irregular. Figure 2-4 shows an example of how characteristic impedance can be determined in simple geometries.

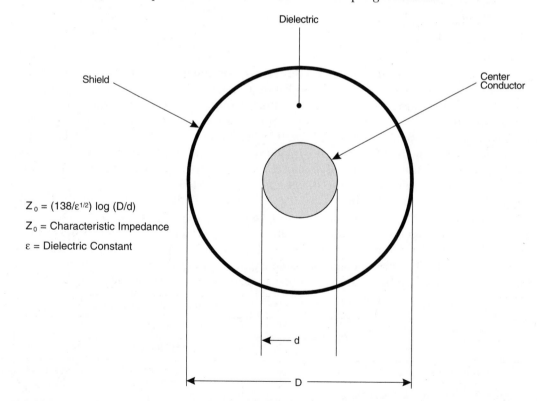

$Z_0 = (138/\varepsilon^{1/2}) \log (D/d)$

Z_0 = Characteristic Impedance

ε = Dielectric Constant

Figure 2-4. Characteristic impedance of coaxial cable

The most efficient transfer of electrical energy occurs in a system where all parts have the same characteristic impedance. If a signal traveling down a transmission path meets a change in impedance, a portion of the energy will be reflected back toward the source. This reflected energy is also noise; that is, it is unwanted energy in the system that can distort signals.

How much energy is reflected depends on the severity of the impedance mismatch. If the mismatch is large in terms of the difference in ohms, a greater amount of energy will be reflected.

A component's electrical dimensions are an important concept in understanding signal transmission. Electrical length is compared to the wavelength of the signal traveling though the components. Figure 2-5 shows the relationship between a signal's frequency and its wavelength. A 10-MHz signal, for example, has a wavelength of 30 meters, while a 100-MHz signal has a wavelength an order of magnitude shorter – 3 meters. If a component is much smaller than the wavelength, it is electrically insignificant. The signal doesn't "see" the component.

It's important to remember that this relationship between the wavelength of the signal and the physical size of components applies only to characteristic impedance and reflections. Crosstalk and attenuation, on the other hand, are not similarly affected by a component's electrical dimensions.

Frequency	Wavelength (meters)
1 kHz	300,000
10 kHz	30,000
100 kHz	3,000
1 MHz	300
10 MHz	30
100 MHz	3
1 GHz	0.3
$f = c/\lambda \quad \lambda = c/f$	

Figure 2-5. Frequency versus wavelength

The significance of a component's size can only be seen in comparison to the wavelength of the signal passing through it. As the signal speed increases, the wavelength decreases and the component becomes electrically larger. Therefore, at very high speeds, even a small component becomes electrically significant.

Notice that energy can reflect in both directions. Consider the situation in Figure 2-6, in which the line has three different impedances. At the first discontinuity, a small portion of the energy is reflected and the rest continues on. At the second, some energy is again reflected. But what happens when this back-reflected energy meets the first discontinuity? Some energy is re-reflected so that it is travelling in the forward direction again.

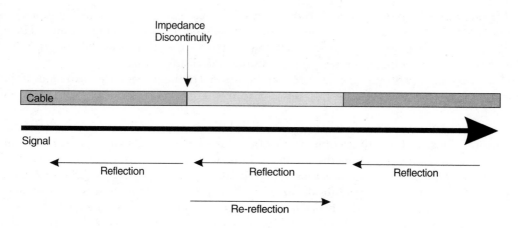

Figure 2-6. Impedance mismatches and reflections

Again, the most important thing to remember about impedance is this:

> *The most efficient transfer of energy occurs when all parts of a circuit have the same characteristic impedance.*

> *Impedance mismatches are of greater concern at higher frequencies than at lower frequencies. The higher the frequency, the more important impedance matching becomes.*

The reason is that at high frequencies, components tend to be electrically shorter and therefore more electrically significant than at lower frequencies.

The task of cable design is to devise cables that transmit signals as noise-free as practical. Once signal transmission quality is achieved, other issues such as cost and ease of use can be addressed.

ATTENUATION

Attenuation is the loss of signal power along the length of the cable. A significant feature of attenuation in copper cable is that it increases with frequency. Attenuation per unit length is greater at 10 MHz than at 1 MHz. Figure 2-7 shows typical attenuation for coaxial and twisted-pair cables.

Attenuation must not become severe enough that the receiver can no longer distinguish the signal from the noise. Attenuation significantly limits practical cable lengths, especially at high speeds where attenuation is more severe.

Frequency (MHz)	Attentuation (dB/100 meters)				
	Thick Coax	Thin Coax	Cat. 3 UTP	Cat. 4 UTP	Cat. 5 UTP
1	0.62	1.41	2.6	2.2	2.0
10	1.70	4.26	9.7	6.9	6.5
20		6.0		10.0	9.3
50	3.94	9.54			
100		13.70			22.0

Note: UTP figures are based on TIA/EIA requirements for horizontal cable.

Figure 2-7. Attenuation for coaxial and UTP cables

COAXIAL CABLE

Coaxial cable offers the best high-frequency performance of copper cables. Figure 2-8 shows the basic structure of a coaxial cable: center conductor, dielectric, shield, and jacket. These components are coaxial: they share the same center axis. As was shown earlier in Figure 2-5, the characteristic impedance of a coaxial cable is determined by three things: diameter of the center conductor, diameter of the shield, and the dielectric constant of the material separating them.

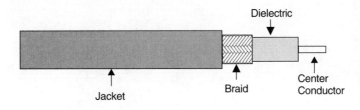

Figure 2-8. Basic construction of coaxial cable

A coaxial cable is a closed transmission line. The shield serves to confine the energy within the cable. While we often think of a wire as carrying electrons through the conductor, it is equally important to think about the electromagnetic energy propagating down the wire. This electromagnetic energy is the magnetic and electric fields generated by the movement of the electrons. These fields travel on the outside of the center conductor, in the dielectric, and on the inside of the outer conductor. This system is called "closed" because the shield serves to keep the energy trapped within the cable.

In a coaxial cable, crosstalk is of very little concern. The shield works well at keeping the energy inside from radiating out or coupling to adjacent cables. Conversely, it protects the signal by keeping outside noise from getting in. Still, the problem remains of noise and signal distortion that results from impedance mismatches. This reflected energy, to a great degree, is also trapped within the cable.

But the shield isn't perfect. Some shields are woven from very thin wires. Such braided shields provide coverage on the order of 85% to 95%. There are still gaps from energy to enter or exit. Other shields are foils, similar to aluminum foils. While these tend to provide 100% coverage, they are thin enough for some of the energy to penetrate.

Two variations of coaxial cable are twinaxial cable and triaxial cable, shown in Figure 2-9. Twinax has two center conductors running parallel within the dielectric. IBM midrange computers use a 93-ohm twinax cable to connect to terminals. Twinax is somewhat similar to the twisted-pair cable we will discuss shortly in that it uses differential transmission.

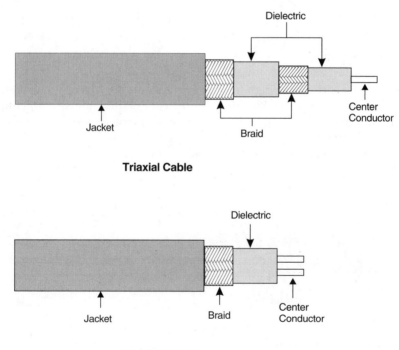

Figure 2-9. Twinax and triax cable

COAXIAL CABLE CHARACTERISTICS

Characteristic Impedance

Most coaxial cables used in LAN applications have a characteristic impedance of 50 ohms. CATV and other video systems typically use 75-ohm cable.

#3

Velocity of Propagation

The speed at which a signal propagates through the cable is determined by the dielectric constant of the dielectric. While electromagnetic energy travels at the speed of light, what we commonly call the speed of light – 300,000 kilometers per second or 186,000 miles per second – is the velocity of light in free space. Electromagnetic energy travels slower in other mediums.

A cable's velocity of propagation is usually expressed as a percentage of the energy free-space speed: 70% or .7c – equivalent to 210,000 kilometers per second or 130,200 miles per second. In terms more practical to premises wiring, this speed translates to 210 meters or 688 feet in 1 microsecond.

Velocity of propagation is handy in several ways. Networks have certain rules about how long a signal can travel along the network. For a given distance, velocity of propagation can determine how long it takes the signal to travel the path. Some network designers will calculate transit times for signals.

Velocity of propagation can also allow you to measure the length of a cable from one end if the other end is unterminated. A signal is injected into the cable. It travels to the other end, where the large impedance mismatch will reflect it back to the source. The round-trip time is measured. If you know the velocity of propagation, you can calculate the cable's length. The same technique can be used to locate cable faults such as breaks: you measure the time it takes energy to reach the break and reflect back. From the velocity of propagation, you can determine the distance (or, even better, have your test instrument do the calculations for you).

Propagation Delay

Closely related to propagation velocity, propagation delay is the time it takes a signal to travel the length of the cable. To calculate signal propagation delay times in a network requires that you know the delays of all components in the system, including hubs, patch cables, and so forth. A network hub, for example, represents a propagation delay. Depending on the hub, its design, and the amount of processing of the signal that must occur, this delay can vary from small to relatively large. Most vendors can supply propagation delay values for their equipment if you need to calculate network delays.

Termination

A coaxial cable must be terminated in its characteristic impedance. Some networks use a bus structure in which neither end of the cable necessarily connects to a device. A special connector called a terminator is placed on the cable end to absorb all the energy reaching it. This prevents reflections. If the cable is not terminated, the cable end represents an infinite impedance that will reflect nearly all the energy. If the shield and center conductor are shorted, the signal will also reverse and travel down the cable as noise. So termination is essential to proper network operation. Always be sure that the terminator has the correct resistance value. Terminating a 50-ohm cable with a 75-ohm terminator still causes reflections.

COAXIAL CABLES FOR NETWORKS

In 1973, the Xerox Corporation began developing a network they named Ethernet. The first prototype was running successfully by 1976 at the Xerox Palo Alto Research Center. It used a large, sturdy, almost unwieldy coaxial cable with a diameter of either 0.405" for plenum applications or 0.375" for nonplenum applications. The cable uses a four-layer shield: braid, foil, braid, foil.

The cable has a 50-ohm (±2 ohms) characteristic impedance and uses N-type connectors. Its large diameter makes the cable stiff and, therefore, hard to work with.

By 1988, a new version of Ethernet had been developed that used a thinner, more flexible, and less expensive coaxial cable. This 50-ohm cable, RG-58A/U, uses a stranded center conductor and a single-layer braided or foil shield. This version of Ethernet became popularly known as thinnet or cheapernet. The earlier version, in turn, was called thicknet. Thinnet cables use BNC connectors.

Broadband cable for carrying video signals uses 75-ohm cables and the same F-connectors that are used in the CATV industry. These cables find little use today in networks except in specialized applications. Digital Equipment Corporation uses 75-ohm cable in its DECnet system to carry video signals.

Still, as Ethernet grew in popularity, network designers searched for additional ways to reduce the cost of cabling. At the same time, IBM introduced its network system called Token Ring. Token Ring networks originally used shielded twisted-pair cable. Soon designers also developed unshielded twisted-pair cable – like standard telephone cable but better. Today, twisted-pair cable is the favored copper cabling system for premises.

Figure 2-10 shows thicknet and thinnet cables.

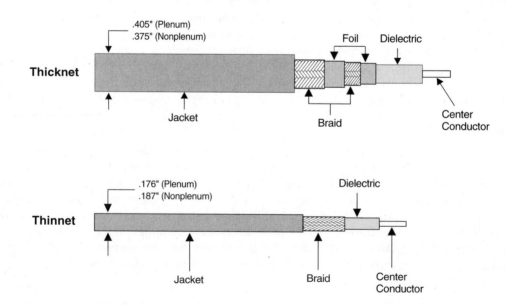

Figure 2-10. Coaxial cables used in Ethernet networks

TWISTED-PAIR CABLES

Twisted-pair cables are so named because each pair of wires is twisted around one another as shown in Figure 2-11. Twisted-pair cables help reduce crosstalk and noise susceptibility in two ways.

First, the twists reduce inductive coupling between pairs. Inductive coupling is caused by the expanding and contracting magnetic fields caused by a signal through the wire. The twists create coupling among the two wires in the pair such that opposing electromagnetic fields are canceled. A greater twist ratio – the number of twists per unit length – brings greater coupling and less chance of crosstalk to nearby pairs.

The optimum twist rate involves a tradeoff between crosstalk and other cable characteristics. More twists lower crosstalk, but increase attenuation, propagation delay, and costs. High-quality Category 5 cables have about one to three twists per inch. For best results, the twist length should vary significantly from pair to pair.

Second, networks using twisted-pair wiring use balanced transmission. The balanced circuit uses two wires to carry the signal. The conductors and all circuits connected to them have the same impedance with respect to ground and other conductors. Ground serves as a reference potential, rather than as a signal return.

Figure 2-11. Twisted-pair cable

Each conductor in a balanced line carries a signal (and noise) that is of equal potential with the other conductor, but of opposite polarity. If one conductor carries a 2.5-V signal, the other conductor carries a -2.5-V signal. The receiver detects only the voltage *difference* between the two conductors – in this case, 5 V. If a +1-V noise is introduced onto each conductor, the resulting signal is 3.5 V and -1.5 V. The receiver still sees a 5-V difference. A perfectly balanced system is difficult to achieve, so some difference in noise levels may appear between the two conductors. Still, the effect is to effectively cancel the noise.

Categories of Twisted-Pair Cable

The telephone industry has used twisted-pair cables for years. It's common to say that networks run over telephone cable. This is wrong for two reasons. First, not all telephone cable is twisted pairs. Quad wiring, found in homes, has four parallel wires, not two twisted pairs. Likewise, silver satin cable, which is flat with a silver jacket, is unsuited for high-speed data communications. Silver satin cable is often used to connect computer modems to wall jacks: it's fine for the slow speeds of modems. Even a fast 28.8-kbps modem requires a cable with a bandwidth of only 3 kHz.

Even twisted-pair cables designed for low-speed analog telephony are unsuitable for high-speed network applications. Telephone cables were designed to carry voices with frequencies of up to only 4 kHz. Because of this low frequency, the twists in the wire can be long and "sloppy." By sloppy, we mean the precision and uniformity of the twists are not of great importance. What is important is that the wires *are* twisted. A telephone twisted-pair cable has about one twist per foot.

By putting more twists per unit length and making them more uniform, cable makers developed cables capable of supporting data rates well above the 4 kHz of telephone cables. Soon, different manufacturers were offering a wide range of better cables under terms like *data grade, high frequency, extended frequency,* and so forth. There was some confusion since these improved cables were offered in various grades, with slightly different specifications from different manufacturers, and each with different terms.

A major distributor, in an attempt to bring order to its offerings, offered twisted-pair cable in four levels of performance. Later a fifth level was added. These cables were termed Level 1, Level 2, Level 3, Level 4, and Level 5. Particularly, Levels 3, 4, and 5 had a rigorous set of specifications governing their performance. Standard-setting bodies, such as UL, IEEE, and TIA/EIA, began building specifications around these levels.

Just as the idea of levels became popular, the distributor decided to trademark the term *Level* in reference to twisted-pair cable. In other words, they wanted to own the term *Level.* The industry quickly adopted the term *Category* to replace *Level.* So today we have Categories 1 through 5 cable. (The trademark was rejected by the U.S. Trademark and Patent Office, incidentally. So the whole problem was much ado about nothing.)

The basic idea of categories of cables is to provide different grades to meet different data rates. But to meet these data rates, other performance characteristics like NEXT must also be specified. In short, here are the frequencies supported by the various cable categories:

Category 1: not rated

Category 2: 64 kHz

Category 3: 16 MHz

Category 4: 20 MHz   # 4

Category 5: 100 MHz

Neither Category 1 nor 2 cables are recognized by TIA/EIA-568A or ISO/IEC 11801 for premises cabling applications. They recommend Category 3 as the minimum grade of UTP.

Category 1 is suited to low-speed analog telephony. But don't conclude that all telephone wiring is Category 1. Telephone companies also use higher grade cables, especially in newer buildings with digital PBXs. The higher grade cables used by different telephone companies differ somewhat in specifications (and many do not exactly match TIA/EIA categories), but they can be used for data communications. If you have existing telephone wiring in a building, it's often best to test it to see what level of performance it can support.

MEGAHERTZ VERSUS MEGABITS

At this point, we should take a short detour to discuss the relationship between the frequency rating of a cable and the data rate of the digital signals carried by the cable. For example, Category 5 cable is rated at 100 MHz, while FDDI has a data rate of 100 Mbps. Are the frequency rating and data rate equivalent? Sometimes yes, sometimes no. The reason is the encoding method used.

A string of digital highs and lows is often unsuitable for transmission over any appreciable distance. Where a digital system uses simple high and low pulses to represent 1s and 0s of binary data, a more complex format is often needed to transmit digital signals over a network. A modulation code is a method for encoding digital data for transmission.

Figure 2-12 shows several popular modulation codes. Each bit of data must occur within its bit period, which is defined by the clock. The clock is a steady string of pulses that provide basic system timing. Some codes are self-clocking and others are not. A self-clocking code means that the clock or timing information is contained within the code. In a non-self-clocking system, this timing information is not present.

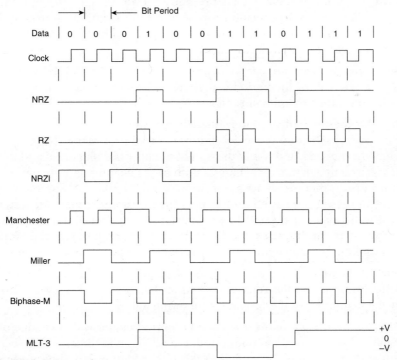

Figure 2-12. Modulation codes

Clock information is important to a receiver. One purpose of a receiver is to rebuild signals to their original state. In order to do so, the receiver must know the timing information. There are three alternatives:

1. The transmitted information can also contain the clock information; in other words, the modulation code is self-clocking.

2. The clock or timing information must be transmitted on another, separate line. This, of course, adds to system complexity by increasing the number of lines from transmitter to receiver.

3. The receiver provides its own timing and does not rely on clock signals from the transmitter.

Modulation Codes

NRZ Code. The *NRZ (nonreturn-to-zero) code* is similar to "normal" digital data. The signal is high for a 1 and low for a 0. For a string of 1s, the signal remains high. For a string of 0s, it remains low. The level changes only when the data level changes.

RZ Code. The *RZ (return to zero) code* remains low for 0s. For a binary 1, the level goes high for one-half of a bit period and returns low for the remainder. For a string of three 1s, for example, the level goes high-low, high-low, high-low in three bit periods.

NRZI Code. In an *NRZI (nonreturn-to-zero inverted) code,* a 0 is represented by a change in level; a 1 is represented by no change in level. Thus, the level will go from high to low or from low to high for each 0. It will remain at its present level for each 1. An important thing to notice here is that there is no firm relationship between the 1s and 0s of data and the highs and lows of the code. A binary 1 can be represented by either a high or a low. So can a binary 0. NRZI is sometimes called NRZ-S (NRZ-space).

Manchester Code. A *Manchester code* uses a level transition in the middle of each bit period. For a binary 1, the first half of the period is high, and the second half is low. For a binary 0, the first half is low, and the second half is high.

Differential Manchester Code. A differential Manchester code also uses a transition in the middle of the bit period. The difference lies in how the transition is interpreted. To represent a 1, the level remains the same as it was at the end of the previous bit period and then changes at midpoint. To represent a 0, the level changes both at the beginning of the bit period and at the end. In other words, there is always a transition in midperiod. A 1 has a transition at the beginning of the bit period; a 0 has no transition at the beginning. The Manchester code also defines two code violations called J and K, which are used for special purposes. A J violation occurs when there is no transition at both the beginning and midpoint of the bit period. A K violation occurs when there is a beginning transition, but no midperiod transition.

Miller Code. In the *Miller code or delay modulation,* each 1 is encoded by a level transition in the middle of the bit period. A 0 is represented either by no change in level following a 1 or by a change at the beginning of the bit period following a 0.

Biphase-M Code. In the *biphase-M code,* each bit period begins with a change of level. For a 1, an additional transition occurs in midperiod. For a 0, no additional change occurs. Thus, a 1 is both high and low during a bit period. A 0 is either high or low (but not both) during the entire bit period.

MLT-3 Code. Unlike other codes, MLT-3 code uses three levels of transmission, rather than two levels. The three levels are -1 V, 0 V and +1 V. Any change in a bit means a change in level. The level stays the same for consecutive 1s and 0s. The movement is upward, then downward as bit changes occur: 0, +1, 0, -1, 0, +1, 0, -1. MLT-3 coding is widely used in high-speed networks using UTP as the cabling medium because the code keeps the frequency content of the signal low to minimize crosstalk and attenuation.

nB/nB Encoding

Most low-speed networks use Manchester or differential Manchester encoding. At higher speeds, Manchester encoding falls from favor because it requires a clock rate twice that of the data rate. A 100-Mbps network requires a 200-MHz clock. NRZI data transmissions have no transitions when all zeros are present, which eliminates self-clocking.

High-speed networks use a group encoding scheme in which data bits are encoded into a data word of longer bit length. For example, the 4B/5B method encodes four data bits into a 5-bit code word. The receiver decodes the 5-bit word into 4 bits. This scheme guarantees that the data never have more than three consecutive 0s. Such encoding requires much less band-width than Manchester encoding. The overhead is only 20%. Thus high-speed networks using 4B/5B, such as FDDI or Fast Ethernet, transmit at a 125-Mbps rate, but the data rate is 100 Mbps. The 25-Mbps difference is due to the fifth bit in the 4B/5B encoding. Even with 4B/5B, the data is still transmitted with one of the encoding schemes in Figure 2-12. Other encoding schemes include the 5B/6B scheme used by 100VG-AnyLAN (described in Chapter 5) and the 8B/10B method used by IBM in its fiber-optic interconnection system for large computers.

Different encoding schemes are more efficient in their use of bandwidth than others. By bandwidth, we mean the comparable *frequency* of the signal. There is not necessarily a one-to-one relationship between data rate (in Mbps) and frequency (in MHz). Figure 2-13 summarizes the relationship between common network encoding schemes, their data rates, and their frequency. Notice that TP-PMD uses MLT-3 encoding with a 125-Mbps transmission rate and a 100-Mbps data rate, but has a frequency of only 31.25 MHz. In other words, TP-PMD is quite efficient, requiring only one-third of the bandwidth available on Category 5 cable. *Bandwidth efficiency* is found by dividing the transmission rate by the frequency requirement (bandwidth used).

Application	Coding Scheme	Data Rate (Mbps)	Transmission Rate (Mbps)	Bandwidth Used (MHz)	Bandwidth Efficiency
Token Ring	Manchester	4	4	4	1
Ethernet	Manchester	10	10	10	1
Token Ring	Manchester	16	16	16	1
100VG-AnyLAN	NRZ	25 (100)*	30 (120)*	15	2
TP-PMD	MLT-3	100	125	31.25	4
ATM	CAP-16	51.84	51.84	25.92	2
ATM	NRZ	155.52	155.52	77.76	2

100VG-AnyLAN has a data transmission rate of 100 Mbps, but breaks the signal into four separate 25-Mbps channels for transmission.

Figure 2-13. Relationship between data rate and frequency for different modulation codes and network types

Even though Fast Ethernet has a transmission rate of 125 Mbps, it requires a cable capable of carrying only 31.25 MHz. But you don't know the relationship between transmission rate and frequency unless you know the encoding scheme used.

The important thing to remember is that the transmission rate and frequency bandwidth of the cable are not always the same.

Carrierless Amplitude and Phase Modulation

The encoding techniques above are digital; that is, they directly use pulses. CAP modulation is related to the techniques used in modems. A frequency is modulated by changing its amplitude and phase. For example, CAP-16 uses 16 different combinations of phase and amplitude to represent bits. Each specific combination represents four bits, as shown in Figure 2-14. This allows high bit rates at low frequencies. For example, 51.8 Mbps ATM, which is used to carry ATM to the desktop, uses CAP-16 encoding. While the CAP-16 systems theoretically require only a 12.96 MHz frequency, the ATM system uses double that frequency or 25.92. This keeps the frequency spectrum under 30 MHz. 30 MHz is a "magic" number because FCC regulations regarding EMI emissions from equipment begin at 30 MHz. Thus keeping the frequency under 30 MHz not only eliminates considerations involved in complying with government regulations, but it also minimizes EMI problems in the first place.

Another CAP scheme is CAP-64, which can represent 6 bits for every discrete combination of phase and amplitude. The number of bits that can be represented is equal to 2 raised to the power of the number of bits. For CAP-16, $2^4 = 16$. In other words, if you have sixteen amplitude/phase combinations, you can represent four bits per combination. Similarly CAP-64 is $2^6 = 64$, so you can have six bits per combination if you have 64 possibilities.

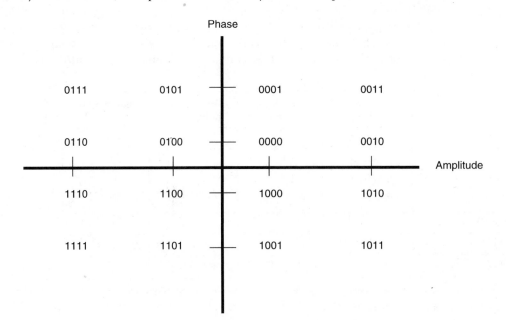

Figure 2-14. CAP-16 encoding

UNSHIELDED TWISTED PAIRS

Unshielded twisted-pair (UTP) cable is the most popular form. The cable categories mentioned earlier are for UTP cable.

The properties of UTP cables have been standardized by a number of interested bodies. The main specification for the cable has been TIA/EIA TSB-36, which defines the electrical and performance characteristics for Categories 3, 4, and 5. The electrical characteristics include characteristic impedance, capacitance, inductance, and NEXT. TSB-36 has been succeeded by TIA/EIA-568A, which covers building cabling. The drawback to TSB-36 is that it defines only cables, while a high-performance cabling plant must take into account the entire *cabling system*. The system's performance is only as good as its weakest component. TIA/EIA-568A defines the cabling system and its performance requirements.

UTP is the favored copper cable for premises wiring for several reasons. It is simple flexible and easy to work with; connectors and related hardware are the familiar modular plugs and jacks long used in the telephone industry. UTP is considered to be inexpensive, well known, and widely available.

The Special Case of Category 5

Even with these advantages, Category 5 cable requires special attention. It can be quite sensitive to incorrect installation. For example, its performance rests on maintaining the twisting of pairs as close to the connector as possible – ½ inch or less. Likewise, clinching a cable tie too tightly when securing cables can also degrade performance. To the installer used to working with Category 2 or 3 cable, which is insensitive to generous untwisting and rugged to abuse, Category 5 cable can be a new experience.

With a premises cabling plant meant to operate at 100 MHz, other interconnection hardware – connectors, patch panels, wall outlets – must also be rated for Category 5 performance. You can't simply go down to the local Builder's Square or Radio Shack and load up on the same components you'd use to wire telephones.

Better grades of cable, of course, mean higher costs. Category 5 components cost more than Category 3 or 4 components. In fact, at slower speeds, you *could* buy some of the hardware at the local Builder's Square.

Twisted-Pair Construction

The basic configuration of UTP cable is a four-pair cable: four individually jacketed twisted-pair cables are held within a single outer jacket as shown in Figure 2-15. The wire size is 24-gauge stranded or solid conductors (although 22-gauge wire is sometimes used). Each conductor's insulation is color coded for easy identification (see Appendix D for color coding information). Four-pair cables are used mainly in the horizontal wiring.

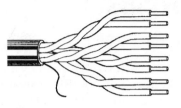

Figure 2-15. Four-pair unshielded twisted-pair cable *(Courtesy of Belden Wire and Cable Company)*

For vertical backbone wiring, a 25-pair cable is often used. A two-pair cable is sometimes used for telephone cables. However, it is recommended that four-pair cable be used in the horizontal subsystem of new buildings to ensure maximum flexibility. For example, a four-pair Category 5 cable is equally suited to analog telephone, digital telephone, and network applications. If only a two-pair wire were installed, the cable would not be capable of being used in some applications. Figure 2-16 shows a cable having many pairs.

Figure 2-16. Multipair cable *(Courtesy of AT&T)*

The most common impedance for UTP is 100 ohms. TIA/EIA-568A recognizes only 100-ohm UTP for buildings. ISO/IEC 11801, on the other hand, also recognizes a 120-ohm UTP, which is popular in Europe (especially France). For balanced cables, the impedance is different from the characteristic impedance.

UTP CHARACTERISTICS

A listing of pertinent UTP characteristics for Categories 3, 4, and 5 cables, as defined in TIA/EIA-568A, is given in Appendix A. Those listed are the properties most often tested during installation and certification of premises cabling.

Four items are of interest here: NEXT, structural return loss, attenuation, and attenuation-to-crosstalk ratio. These deserve some discussion.

NEXT and Power-Sum NEXT Losses

Figure 2-17 shows graphically the NEXT loss requirements for horizontal cable. Appendix A supplies further information on how NEXT is calculated.

The NEXT loss requirements for Category 4 cable are 15 dB tighter than for Category 3. NEXT levels must be 15 dB lower. For Category 5 cable, NEXT is 8 dB lower than Category 4 cable and 23 dB lower than Category 3 cable. Notice that we use the term *lower* even though the dB value is a larger number. Remember that a high NEXT loss value means that the crosstalk is lower compared to the signal. A higher NEXT loss value means lower crosstalk.

If you actually measure NEXT across the entire frequency, you'll see that crosstalk jumps around with changes in frequency. Figure 2-18 shows measured NEXT loss for a Category 5 cable. Notice that it does not increase or decrease linearly with frequency. It shows sharp valleys across the spectrum. The figure also shows smooth curves that relate to 568A requirements.

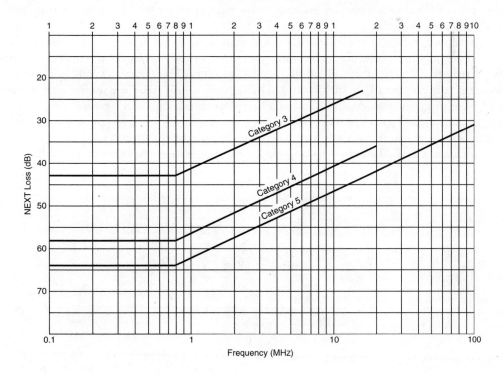

Figure 2-17. UTP NEXT loss requirements

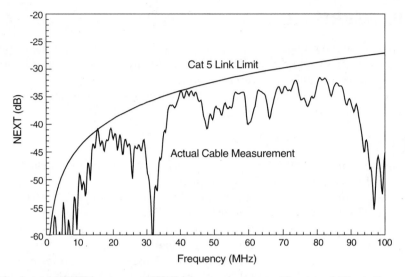

Figure 2-18. Actual NEXT loss versus NEXT loss requirements *(Courtesy of Microtest)*

For horizontal cable, straight NEXT loss measurements on the cable can be made. A driven pair is energized and the crosstalk on the quiet pair is measured at the near end. Only two pairs of cable are involved in each measurement. The measurement for the worse-case pair is the one used in evaluating the cable.

For backbone cables, a power-sum measurement is used. While the NEXT loss measurement involves only two pairs, the active and quiet ones, in the real world the quiet pair could receive crosstalk coupled from all other pairs. Consider 25-pair cable. While the coupling between any pair can meet Category 5 standards on a pair-to-pair basis, the total energy coupled from 24 pairs onto one pair could be too much. The reason is that in a typical Category 5 horizontal cable, only two pairs are used (even in a four-pair cable). In a 25-pair backbone cable, many pairs can be active at one time. To ensure transmission integrity, the crosstalk on each pair must be measured as if all pairs were energized. Figure 2-19 shows power-sum crosstalk.

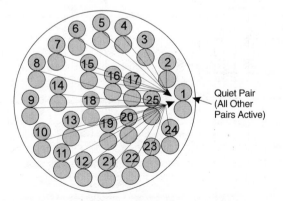

Figure 2-19. Power-sum crosstalk accounts for coupling simultaneously from all other pairs in the cable

In making measurements for power-sum crosstalk, you don't have to energize all 24 pairs at the same time. Rather, each pair is energized individually and the crosstalk on the single quiet line is measured. For example, to find the power-sum crosstalk on pair 1, you would drive pair 2 with a signal and measure the NEXT on pair 1. Then you would drive pair 3 and measure the NEXT on pair 1. Then drive pair 4 and measure NEXT on pair 1. This would continue to pair 25. The power-sum crosstalk is then calculated:

$$NEXT_{PS} = (NEXT_{PR2\text{-}1}{}^2 + NEXT_{PR3\text{-}1}{}^2 \ldots + NEXT_{PR25\text{-}1}{}^2)^{1/2}$$

Next, perform the same procedure on the other 24 pairs. Measure NEXT coupled onto pair 2, then pair 3, and so forth. The power-sum calculations are typically 3 to 6 dB worse than a straight pair-to-pair calculation. The 6-dB difference is important; it's the difference between NEXT performance of Category 4 and 5 cable. A cable that barely meets Category 5 pair-to-pair NEXT will fail the power-sum Category 5 test and will barely meet Category 4 requirements.

Unfortunately, there exists no practical method of testing 25-pair UTP in the field. The cable manufacturer tests the cable in the laboratory using a network analyzer. This equipment is expensive, requires careful setup and calibration, and is unsuitable for general field use.

Structural Return Loss

Structural return loss (SRL) is a measure of the uniformity of a cable's characteristic impedance. SRL is affected by the design and manufacturing of the cable, since these result in variations in the structure of the cable: variations in twisting, differences in the dielectric, and variations in separation distance between conductors that make the cable less than a perfect cable. SRL is measured in decibels, with a higher number meaning a better cable. Again, the important SRL is the worst one in the cable. While TIA/EIA-568A provides a formula for determining SRL at any frequency, Figure 2-20 shows SRL requirements. Notice that the higher the cable category, the better the SRL is (i.e., the more uniform the impedance). At the same time, for each type of cable, SRL decreases at higher frequencies.

SRL can contribute noise, just as crosstalk does. NEXT is usually the main source of noise. What the SRL specification does is ensure that the SRL-induced noise is never more than 20% of the *total* noise at the receiver.

Attenuation-to-Crosstalk Ratio

Think for a moment about signals and crosstalk. If a given level of crosstalk is coupled onto the cable, its relationship to the signal depends on the signal's power. If the signal is strong, then the crosstalk is only a small percentage of the signal. If the signal is weak, or highly attenuated after traveling down the cable, then the crosstalk is a greater percentage of the signal. If the coupled crosstalk is .5 V and the signal is 5 V, the noise is 10% of the signal. If the signal is attenuated to 2.5 V, the noise is now 20% of the signal.

The attenuation-to-crosstalk ratio is a figure of merit that accounts for the relative strength of the signal. Figure 2-21 shows the relationship between NEXT and attenuation. The distance that separates them defines the ACR. ACR generally decreases at higher frequencies. At 10 MHz for Category 5 cable, NEXT loss is 47 dB and attenuation is 6.5 dB. The ACR is 31.5 dB. At 100 MHz, where NEXT is lowered to 32 dB and attenuation increases to 22 dB, the ACR is reduced to 11 dB.

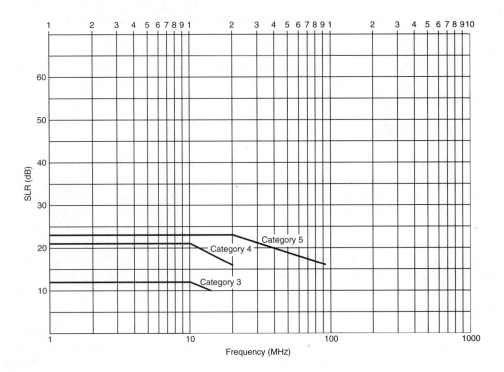

Figure 2-20. UTP SRL requirements

To improve ACR, you can either reduce the crosstalk or lower the attenuation – or preferably do both.

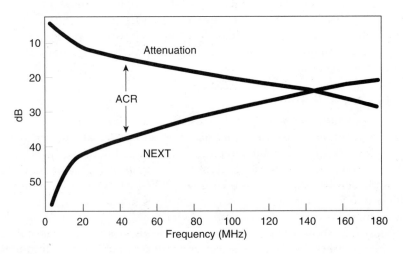

Figure 2-21. ACR

SHIELDED TWISTED-PAIR CABLE

Shielded twisted-pair (STP) cable was popularized by IBM in its Token Ring network. The most commonly used version of STP is called Type 1 cable. IBM created a cabling system, summarized in Figure 2-22, for use with their systems in premises wiring applications. While the system has not caught on as a widely accepted standard, Type 1 cable remains popular as the predominant type of STP.

Name	Type	Impedance (ohms)	Description
Type 1/1A 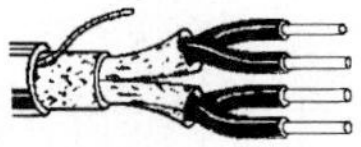	STP	150	Two individually shielded solid-conductor, 22 AWG twisted pairs surrounded by an outer braid shield. Both pairs are suited for data transmission.
Type 2/2A 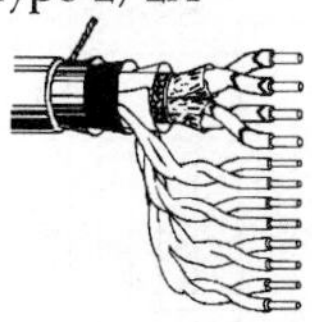	STP/UTP	150 (STP)	Two solid-conductor, 22 AWG twisted pairs surrounded by an outer braid shield. Four solid-conductor, 22 AWG twisted pairs outside the braid for telephone use.
Type 3 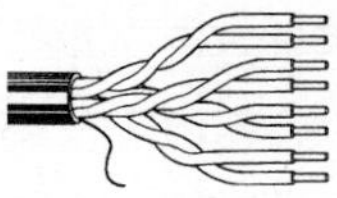	UTP	100	Four solid-conductor 24 AWG twisted pairs, unshielded. Similar to Category 2; not recommended for network applications.
Type 6/6A 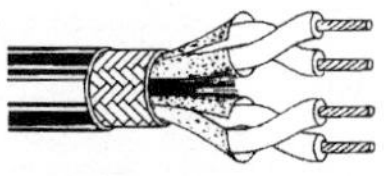	STP	150	Two stranded-conductor, 26 AWG twisted pairs surrounded by an outer braid. Similar to Type 1 except for wire gauge. Used for short patch cords.
Type 9/9A 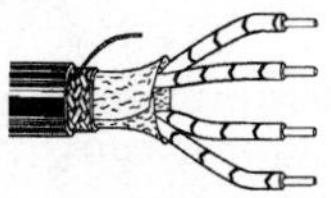	STP	150	A plenum-rated cable with two 26 AWG twisted pairs surrounded by an outer shield.

Figure 2-22. IBM cabling system *(Courtesy of Belden Wire and Cable Company)*

Type 1 cable contains two shielded cable pairs within an outer shield. The original Type 1 cable was rated to 20 MHz; an improved version – Type 1A – is rated to 300 MHz.

STP cable is quite attractive from a performance viewpoint. It allows longer distances and higher speeds than UTP. But UTP is more popular because STP is more expensive and more difficult to work with.

For maximum shielding effectiveness, both ends of a shielded cable should be grounded through a shielded interconnection. One purpose of the shield is to conduct noise to ground. If only one end is grounded, the shield becomes a long antenna, effective only at frequencies for which its length is less than ⅛ of the wavelength. If the shield is not grounded at either end, then it has little effect on noise.

Figure 2-23 shows NEXT values for 150-ohm STP cables.

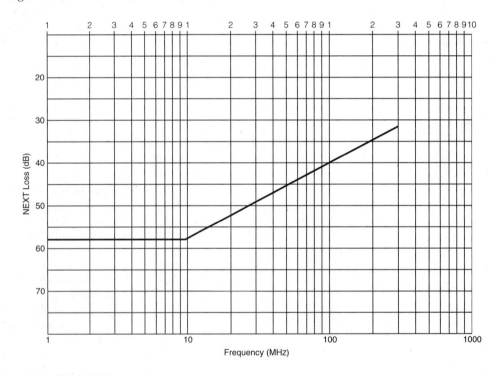

Figure 2-23. STP NEXT loss requirements

SCREENED UTP

Screened UTP is really shielded UTP. But since shielded unshielded twisted pairs is a contradiction of terms, it's called screened UTP (sUTP), screened twisted pair (scTP), or foil twisted pair (FTP). Standard Category 3, 4, and 5 cables are available in shielded versions. As shown in Figure 2-24, the shield is typically a foil shield, which is not as effective as the braided shield used on STP and coaxial cable. Screened UTP cables share the same electrical characteristics as regular UTP, but offer greater immunity to outside noise. Unlike STP, which has each pair individually shielded from each other to reduce NEXT, sUTP has only a single shield around all pairs. The shield does not reduce NEXT; it only protects against outside noise.

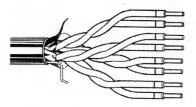

Figure 2-24. Screened UTP *(Courtesy of Belden Wire and Cable Company)*

CABLE LENGTHS

The rule of thumb for total horizontal cabling is 100 meters. Of this TIA/EIA-568A recommends 90 meters from the telecommunications closet to the work area outlet. In addition, it allows 6 meters total for jumper cables, patch cords, and crossconnects inside the telecommunications closet and 3 meters for cable in the work area (from the outlet to the computer). This amounts to 99 meters total. The standard then waffles a bit and allows 10 meters for cable in the telecommunications closet and work area. But the 1-meter difference amounts to a quibble.

More latitude is given for backbone cables. For Categories 3, 4, and 5 UTP and Type 1 STP, the distance is 90 meters, depending on the bandwidth of the signal. Figure 2-25 summarizes distances for the backbone. Notice that Categories 3, 4, and 5 distances depend on the frequency they will be used at. If the frequency falls below the range given, the cable can be treated as a voice-grade application. Below 5 MHz, all UTP cables can be run 800 meters. At 10 MHz, Category 5 cable can still be run 800 meters, but Categories 3 and 4 are restricted to 90 meters.

The distance a cable can be run in the backbone depends on the signal frequency it carries.

FLAMMABILITY RATINGS

Cables are protected by an outer jacket. The jacket protects against abrasion, oil, solvents, and so forth. The jacket usually defines the cable's duty and flammability rating. Heavy-duty cables have thicker, tougher jackets than light-duty cables. Equally important in a building is the cable's flammability rating. The National Electrical Code establishes flame ratings for cables, while Underwriter's Laboratories has developed procedures for testing cables. The NEC requires that cables that run through plenums (the air-handling space between walls, under floors, and above drop ceilings) must either be run in fireproof conduits or be constructed of

Cable	Backbone Cables: Main Cross Connect to Telecommunications Room	
	Frequency	Distance
UTP	Voice grade	800 m
Cat. 3 UTP	5–16 MHz	90 m
Cat. 4 UTP	10 MHz–20MHz	90 m
Cat. 5 UTP	20–100 MHz	90 m
STP	20–300 MHz	90 m
62.5/125 Fiber	–	2000 m
Single-Mode Fiber	–	3000 m

Figure 2-25. Backbone cable distances

low-smoke and fire-retardant materials. For building use, there are three categories of cables depending on the cable type:

Plenum cables can be installed in plenums without the use of conduit.

Riser cables can be used in vertical passages connecting one floor to another.

General-use cables can be used for general installations. They cannot be used in riser or plenum applications without fireproof conduits. These cables can be used in offices spaces – to connect from a wall jack to a computer, for example.

The NEC lists cables by a short acronym describing the cable classification and a final single-letter suffix giving the flammability rating. Figure 2-26 summarizes the classifications for cables discussed in this book. Most cables have their classification printed on the jacket.

Cable Type	Classification		
	Plenum	Riser	Commercial
Communication	CMP	CMR	CM
Fiber with no metallic conductors*	OFNP	OFNR	OFN
Fiber with metallic conductors*	OFCP	OFCR	OFC

Metallic conductors refer to any part of the cable that can conduct electricity, including strength members. The metallic conductor does not have to be intentionally intended to carry electricity or electrical signals. See Chapter 3 for more information about fiber-optic cables.

Figure 2-26. Cable classifications and flammability ratings

CHAPTER 3

Fiber Optics

Fiber-optic technology is significantly different from copper: it uses light transmitted through hair-thin fibers. Compared to copper cable, the fiber-optic cable offers higher bandwidth and lower losses, allowing higher data rates over longer distances.

THE ADVANTAGES OF FIBER

Fiber optics offer distinct advantages over traditional copper media.

Information-Carrying Capacity

Fiber offers bandwidth well in excess of that required by today's network applications. The $62.5/125$-μm fiber recommended for building use has a minimum bandwidth of 160 MHz-km. Because a fiber's bandwidth scales with distance, the bandwidth at 100 meters is over 1.5 Gbps or 5 Gbps (depending on the wavelength of light transmitted). In comparison, Category 5 cable is specified only to 100 MHz over the same 100 meters.

With high-performance single-mode cable used by the telephone industry for long-distance telecommunications, the bandwidth is essentially infinite. That is, the information-carrying capacity of the fiber far exceeds the ability of today's electronics to exploit it.

Low Loss

An optical fiber offers low power loss. Low loss permits longer transmission distances. Again, the comparison with copper is important: in a network, the longest recommended copper distance is 100 meters; with fiber it is 2000 meters or more.

A drawback of copper cable is that loss increases with the *signal* frequency. Attenuation in twisted-pair cable is higher at 100 MHz than at 10 MHz. This means high data rates tend to increase power loss and decrease practical transmission distances. With fiber, loss does not change with the *signal* frequency. Attenuation does change with the frequency of the light transmitted through it, but not for the data rate. Thus, a 10-MHz and 100-MHz signal are attenuated alike in the fiber. Figure 3-1 shows the relationship between signal frequency and attenuation for both copper and fiber cables.

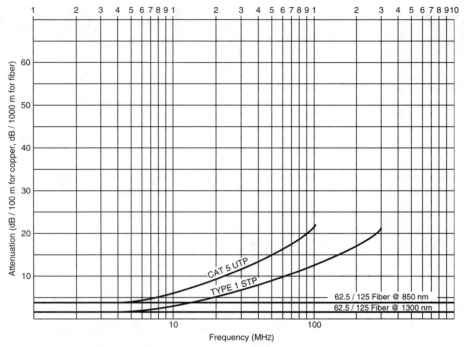

Figure 3-1. Attenuation versus frequency for fiber, UTP, and STP

Electromagnetic Immunity

By some estimates, 50 to 75% of all copper-based network outages are caused by cabling and cabling-related hardware. Crosstalk, impedance mismatches, and EMI susceptibility are major factors in noise and errors in copper system. What's more, such problems can increase with incorrectly installed Category 5 cable, which is sensitive to poor installation.

Because a fiber is a dielectric, it is immune to electromagnetic interference. It does not cause crosstalk, which is a main limitation of twisted-pair cable. What's more it can be run in electrically noisy environments, such as a factory floor, without concern, since noise will not affect the fiber. There's no concern with proximity to noise sources like power lines or fluorescent lights. In short, fiber is inherently more reliable than copper.

Light Weight

Fiber-optic cable weighs less than comparable copper cable. A two-fiber cable is 20% to over 50% lighter than a comparable four-pair Category 5 cable. Lighter weight makes fiber easier to install. Typical weights for 1000 feet of different cables are:

2- fiber cable:	11 lb
12-fiber cable:	33 lb
4-pair category 5 UTP:	25 lb
25-pair backbone UTP:	93 lb
10BASE-2 coax:	24 lb

Smaller Size

Fiber-optic cable is typically smaller than the copper cables it replaces. Again, relative to Category 5 twisted-pair cable, an optical fiber takes up about 15% less space.

Safety

Since the fiber is a dielectric, it does not present a spark hazard, nor does it attract lightning. What's more, cables are available in the same flammability ratings as copper counterparts to meet code requirements in buildings.

Security

Optical fibers are quite difficult to tap. Since they do not radiate electromagnetic energy, emissions cannot be intercepted. And physically tapping the fiber takes great skill to do undetected. Thus the fiber is the most secure medium available for carrying sensitive data.

FIBERS AND CABLES

Fiber optics is a technology in which signals are converted from electrical into optical signals, transmitted through a thin glass fiber, and re-converted into electrical signals. As shown in Figure 3-2, the basic optical fiber consists of three concentric layers differing in optical properties:

Core: the inner light-carrying member.

Cladding: the middle layer, which serves to confine the light to the core.

Buffer: the outer layer, which serves as a "shock absorber" to protect the core and cladding from damage.

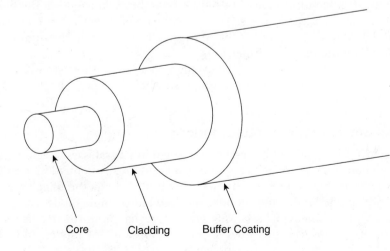

Core   Cladding   Buffer Coating

Figure 3-2. Parts of an optical fiber

Total Internal Reflection

A basic understanding of how light travels through the optical fiber is essential to understanding many of the issues involving fiber properties. Light travels by a principle known as *total internal reflection.*

Light injected into the core and striking the core-to-cladding interface at an angle greater than the critical angle will be reflected back into the core. Since angles of incidence and reflection are equal, the light ray continues to zig-zag down the length of the fiber. The light is trapped within the core. Light striking the interface at less than the critical angle passes into the cladding and is lost. Figure 3-3 shows light traveling through the fiber by total internal reflection.

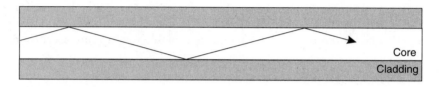

Figure 3-3. Total internal reflection

The rays of light do not travel randomly. They are channeled into modes, which are possible paths for a light ray traveling down the fiber. A fiber can support as few as one mode and as many as tens of thousands of modes. While we are not normally interested in modes *per se,* the number of modes in a fiber is significant because it helps determine the fiber's bandwidth. More modes typically mean lower bandwidth. The reason is dispersion.

As a pulse of light travels through the fiber, it spreads out in time. While there are several reasons for such dispersion, two are of principal concern. The first is modal dispersion, which is caused by different path lengths followed by light rays as they bounce down the fiber. Some rays follow a more direct route than others. The second type of dispersion is material dispersion: different wavelengths of light travel at different speeds. By limiting the number of wavelengths of light, you limit the material dispersion.

Dispersion limits the bandwidth of the fiber. At high data rates, dispersion will allow pulses to overlap so that the receiver can no longer distinguish where one pulse begins and another one ends.

Types of Fibers: Single Mode or Multimode?

The basic structure of an optical fiber can be modified in several ways to achieve different signal transmission characteristics. Modifications affect bandwidth, attenuation, and the practicalities of coupling light into and out of the fiber. Figure 3-4 shows the basic types of fiber: step-index multimode, graded-index multimode, and step-index single mode.

In the simplest optical fiber, the relatively large core has uniform optical properties. Termed a *step-index multimode fiber,* this fiber supports thousands of modes and offers the highest dispersion – and hence the lowest bandwidth.

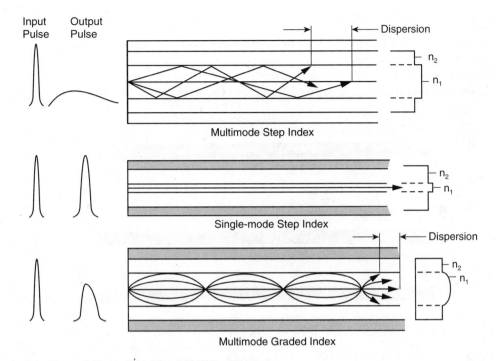

Figure 3-4. Type of fibers *(Courtesy of AMP Incorporated)*

By varying the optical properties of the core, the *graded-index multimode fiber* reduces dispersion and increases bandwidth. Grading makes light following longer paths travel slightly faster than light following shorter paths. The net result is that the light does not spread out nearly as much. Nearly all multimode fibers used in networking and data communications have a graded-index core.

But the ultimate in high-bandwidth, low-loss performance is a *single-mode fiber.* Here the core is so small that only a single mode of light is supported. The bandwidth of a single-mode fiber far surpasses the capabilities of today's electronics. Not only can the fiber support speeds tens of gigabits per second, it can carry many gigabit channels simultaneously. This is done by having each channel be a different wavelength of light. The wavelengths do not interfere with one another. Single-mode fiber is the preferred medium for long-distance telecommunications. It finds use in networks for interbuilding runs and will eventually become popular for high-speed backbones.

The most popular fiber for networking is the 62.5/125-µm multimode fiber. The numbers mean that the core diameter is 62.5 µm and the cladding is 125 µm. Other common sizes are 50/125 µm (which is popular in Europe) and 100/140 µm (which is part of the IBM cabling system).

The quick summary:

> *Graded-index multimode fibers – the 62.5/125-µm size in particular – are the preferred fibers for horizontal cable and most backbone applications.*

> *Single-mode fibers, by virtue of their immense bandwidth and long transmission capabilities, are for applications requiring their additional performance.*

Numerical Aperture (NA)

The NA of the fiber defines which light will be propagated and which will not. As shown in Figure 3-5, NA defines the light-gather ability of the fiber. Imagine a cone coming from the core. Light entering the core from within this cone will be propagated by total internal reflection. Light entering from outside the cone will not be propagated.

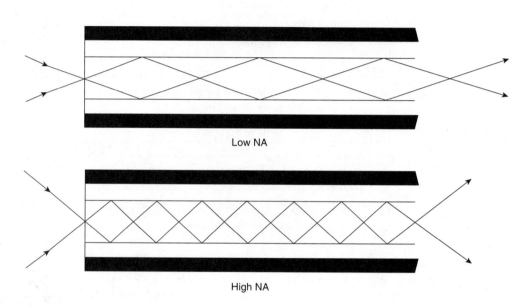

Low NA

High NA

Figure 3-5. Numerical aperture *(Courtesy of AMP Incorporated)*

NA has an important consequence. A large NA makes it easier to inject more light into a fiber, while a small NA tends to give the fiber a higher bandwidth. A large NA allows greater modal dispersion by allowing more modes for the light to travel in. A smaller NA reduces dispersion by limiting the number of modes.

> *The higher the NA and core diameter, the lower the bandwidth of the fiber. Conversely, the lower the NA and core diameter, the higher the bandwidth – and the more difficult it is to inject light into the fiber.*

Attenuation

Attenuation is loss of power. During transit, light pulses lose some of their energy. Attenuation for a fiber is specified in decibels per kilometer (dB/km). For commercially available fibers, attenuation ranges from under 1 dB/km for single-mode fibers to 1000 dB/km for large-core plastic fibers.

Attenuation varies with the wavelength of light. There are three low-loss "windows" of interest: 850 nm, 1300 nm, and 1550 nm. The 850-nm window is perhaps the most widely used, since it was the first developed, and optical devices such as LEDs operating at 850 nm are inexpensive and plentiful. The 1300-nm windows offer lower loss, but at a modest increase in cost for LEDs. The 1550-nm window today is mainly of interest to long-distance telecommunications applications. Figure 3-6 shows attenuation as a function of wavelength for multimode and single-mode fibers used in premises cabling applications. You can see the low-loss points at 850 and 1300 nm.

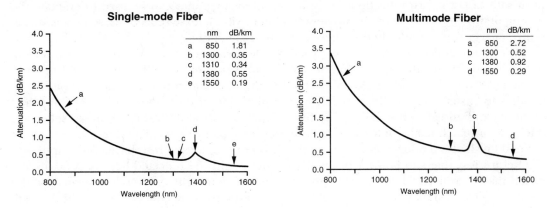

Figure 3-6. Attenuation versus wavelength *(Courtesy of Corning Inc.)*

Bandwidth

Fiber bandwidth is given in MHz-km. Bandwidth scales with distance: if you halve the distance, you double the bandwidth. If you double the distance, you half the bandwidth. What does this mean in premises cabling? For a 100-meter run (as allowed for twisted-pair cable), the bandwidth for 62.5/125-μm fiber is 1600 MHz at 850 nm and 5000 MHz at 1300 nm. For the 2-km spans allowed for most fiber networks, bandwidth is 80 MHz at 850 nm and 250 MHz at 1300 nm.

The discussion of bandwidth efficiency discussed in the last chapter for copper cable does not hold true for fiber-optic cable. The relationship between frequency and bit rate is much closer and can be considered 1:1 in most cases. The fiber has so much available bandwidth that sophisticated and complex coding schemes like MLT-3 aren't needed.

Figure 3-7 summarizes the main characteristics of common fibers.

Fiber Type	Attenuation (Max., dB/km)		Bandwidth (Min., MHz-km)		NA
	850 nm	1300 nm	850 nm	1300 nm	
Single Mode	–	1.0	–	See note	.1
50/125	3.5	2.0	400	400	.20
62.5/125	3.5	1.5	160	500	.275
100/140	5.0	4.0	100	200	.29

Note: The bandwidth of a single-mode fiber is essentially infinite in that it surpasses the abilities of today's electronics to exploit its capabilities.

Figure 3-7. Common fiber characteristics

Return Loss

Just as a signal in copper cable can reflect backward from changes in impedance, discontinuities in an optical fiber can also cause reflections. The greatest source of reflections is at fiber-to-fiber interconnections, especially if the fibers are separated by an air gap. This reflected energy can be propagated back to the transmitter, where it can interfere with operation of a laser source. LEDs, on the other hand, are not adversely affected by this backscattered light. Even with LEDs, return reflections are still measured since they can be used to evaluate the mating of two connectors. An air gap between fibers will increase reflections, so by measuring return energy you can tell whether there is a gap between fibers.

The energy reflected must be minimized. While there are several approaches to reducing the reflected energy, the most common are the physical contact (PC) and angled physical contact (APC) end finishes. In the PC end finish, the fiber and connector end are polished to a radius to ensure that the fibers touch on the high side of the radius. In the APC finish, a slight angle is added to the radius. The fibers touch significantly and this reduces reflection due to a mismatch in the index of refraction between glass and air. The radius and angle of the end finish also tend to reflect light so that it is not propagated by internal reflection.

Return loss refers to how far below the incident light the reflected light is. A 30-dB loss means that the reflected light is 99.9% less than the incident light. A 50-dB loss means that the light is 99.999% less than the incident light. In practical terms, imagine a 50-µW optical signal. The reflected energy would be 0.5 µW for a 30-dB return loss and only 0.005 µW for a 50-dB return loss. Therefore, a higher return loss figure is better.

The TIA/EIA-568A requirements for return loss in connectors are as follows:

Multimode fiber: 20 dB or greater

Single-mode fiber: 26 dB or greater

Other applications, notably high-speed telecommunications, have much higher return loss requirements for single-mode fiber – as high as 67 dB.

More About Cable Distances and Bandwidth

In the last chapter, Figure 2-25 shows that the distances recommended by TIA/EIA-568A are 90 meters for horizontal cable, 2000 meters for multimode backbone cable, and 3000 meters for single-mode backbone cable. The standard also recognizes the conservative nature of these estimates. It notes, for example, that single-mode cables are suited for up to 60 km on the backbone.

Figure 3-8 shows the relationship between distance and bandwidth for fibers in network applications. The graph is based on 62.5/125 fiber driven by an LED operating at 1300 nm. The important thing to notice is that at 100 meters, the fiber's bandwidth exceeds 1 GHz. Even at 2 km, the bandwidth exceeds 200 MHz.

Figure 3-9 presents the same relationship for a single-mode fiber and laser at 1310 nm. At 100 meters, the bandwidth is 333 GHz! Even at 4 km, the bandwidth is over 8 GHz. As we mentioned, the single-mode fiber can support backbone distances of 60 km at 100 MHz.

The distance limitations placed on fiber, then, are artificial ones. Many observers feel that the 90-meter limit on horizontal runs unnecessarily constrains fiber. For example, network specifications routinely allow spans of between 1 to 4 km on a fiber link, yet building cabling

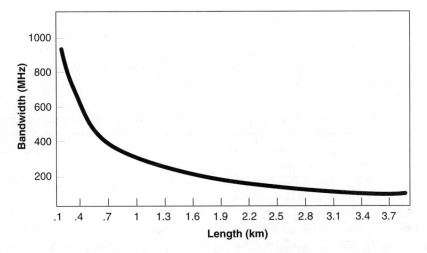

Figure 3-8. Bandwidth-distance curve for 62.5/125-μm multimode fiber at 1300 nm

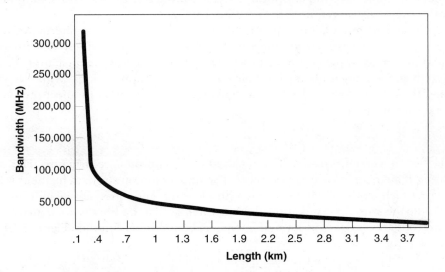

Figure 3-9. Bandwidth-distance curve for single-mode fiber at 1310 nm

requirements shorten the length to 90 meters. While the point is valid, the fact remains the TIA/EIA-568A and ISO/IEC 11801 limit horizontal cable runs.

The TIA/EIA is considering a proposal to allow 300-hundred-meter runs in a combination backbone/horizontal configuration. Current TIA/EIA requirements call for a cross connect to separate the horizontal cable from the vertical cable. The new proposal eliminates the cross connect requirement and allows 300 meters from equipment (such as a network hub) to the work area. This arrangement is commonly called a *homerun* configuration. The 300-meter distance was chosen to ensure that multimode fiber still has enough bandwidth to meet the 622-Mbps speed of ATM.

The 300-meter homerun configuration will most likely be issued as a Technical Service Bulletin (TSB) initially and then be incorporated into a future revision of TIA/EIA-568A. As a TSB, it will become an accepted practice for building cabling. The homerun rules, by making better use of the characteristics of fiber, make fiber a more attractive option.

Plastic Optical Fibers

So far we've been discussing fibers made of glass. They can also be constructed from plastic. Plastic optical fibers have traditionally been step-index fibers with a large core diameter. Plastic fibers have very high attenuation of 200 to 250 dB/km and a low bandwidth of about 5 MHz/km, making them generally unsuitable for network applications. Plastic fibers have large diameters – 250, 500, 750, and 1000 μm are most common – which makes coupling light into the fiber easy. Plastic fibers have found use in short-run, low-bandwidth applications – including digital audio and automotive.

Due to advances in plastic fiber technology, fiber, connectors, and active devices are relatively inexpensive. The fibers operate in the visible light region around 680 nm and inexpensive LEDs and detectors can be used. In addition, connectors are much less costly connectors for glass fibers. In short, step-index optical fibers are attractive from a cost standpoint, but don't offer the performance required for premises cabling.

Recently-developed graded-index plastic fibers may offer a low-cost alternative to glass fibers. Over the 100-meter runs recommended for horizontal cabling, these new fibers have demonstrated the ability to support rates to 1 GHz. A consortium of manufacturers has been formed to study graded-index fibers and their uses in various applications. While to date no systems using these fibers are available, plastic fiber has the potential to affect the premises cabling market significantly.

Cables

The fiber, of course, must be *cabled* or enclosed within a protective structure. This usually includes strength members and an outer jacket. The most common strength member is Kevlar Aramid yarn, which adds mechanical strength. During and after installation, strength members handle the tensile stresses applied to the cable so that the fiber is not damaged. Steel and fiberglass rods are also used as strength members in multifiber bundles. Figure 3-10 shows typical fiber constructions.

The first layer of protection is the buffer, available in loose and tight styles. The tight buffer construction is a plastic layer applied directly to the fiber coating. This buffer is typically 250 or 900 μm thick. Widely used for indoor applications, tight-buffer cables provide great crush and impact resistance, but lower isolation of the fiber from extreme changes in temperature.

The breakout cable is a special form of tight buffered cable – cables within a cable. It consists of several individual tight-buffered cables (fiber, buffer, strength member, and outer jacket) with an outer jacket and additional strength members. Useful for both riser and horizontal applications, breakout cables can simplify routing and installation of connectors.

In the loose-buffer cable, the fiber lies within a plastic tube with an inside diameter many times that of the fiber. A tube can hold more than one fiber. The tube is usually filled with a gel material that serves to keep water out. Since the fiber floats within the tube, it is isolated from external stresses, including expansions and contractions of the cable due to temperature extremes. Widely used in outdoor applications, loose-buffer cables tend to have a larger bend radius, larger diameter, higher tensile strength, and lower crush and impact resistance.

Most multifiber cables have fibers arranged in a circular manner; Figure 3-11 shows a representative multifiber breakout cable. AT&T, for one, offers a ribbon style. Each ribbon has up to 12 fibers in parallel, with up to 12 ribbons (144 fibers) in a cable. Figure 3-12 shows a ribbon cable.

Duplex cables are used for patch cables and many horizontal cables. For backbone and riser cables, cables with 12 or more fibers are recommended.

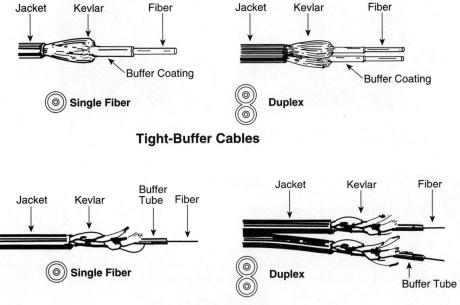

Figure 3-10. Common cable construction *(Courtesy of Belden Wire and Cable Company)*

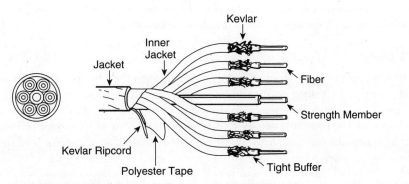

Figure 3-11. Multifiber breakout cables *(Courtesy of Belden Wire and Cable Company)*

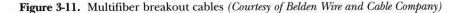

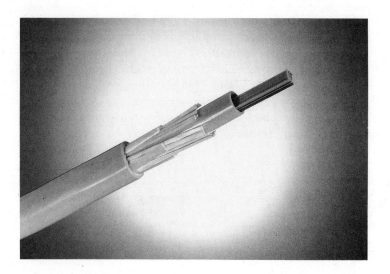

Figure 3-12. Ribbon cable with 144 fibers in 12 ribbons *(Courtesy of AT&T)*

TRANSMITTERS AND RECEIVERS

Getting the Information In and Out

Source and detectors are the fundamental element of the electro-optic interface. The source is either an LED or laser that converts the electrical signal into an optical signal and injects the light into the fiber. The detector is a photodiode that performs the opposite task: it accepts the light from the fiber and converts it back into an electrical signal.

LASERS AND LEDS

The choice of LED or laser for a transmitter depends on the application requirements. Both are small semiconductor chips that emit light when current passes through them. The significant differences between an LED and laser greatly affects the performance of an optical link.

The following are important characteristics of the optical source.

Output Power

How much optical power does the source couple into the fiber? Both LEDs and lasers used in network/premises cabling applications couple tens of microwatts of optical power into a fiber. The amount coupled from a given device depends on the fiber's core diameter and NA. The light from an LED spreads out in a wide pattern so that only a small portion of the total light emitted is actually coupled into the fiber. A fiber with a large core or a higher NA gathers more of this light. For example, an LED might couple 45 μW into a 62.5/125 fiber, but only 35 μW into a 50/125 fiber. The difference is due to the larger diameter and NA of the 62.5/125 cable.

A single-mode fiber requires a laser source. Unlike an LED, a laser produces a very narrow beam of light matched to the small core of the fiber. While lasers used in premises cabling applications couple roughly the same amount of light into a multimode fiber as an LED, they couple much more light into a single-mode fiber. This is because of their narrow beam that more closely matches the small core of the single-mode fiber. The efficiencies of laser sources and single-mode fibers permit longer transmission distances. In an FDDI network, for example, the span between stations is 2 km for multimode fiber and 40 km for single-mode fibers.

An interesting development is the adaption of lasers used in CD players for use in fiber-optic systems. Although CDs use lasers operating at 780 nm, the lasers have been also modified to emit at 850 nm. Both wavelengths are used in fiber optics. While the main attraction of CD lasers is their low cost, their performance characteristics make them well suited to optical communications. They are fast, have a narrow, coherent spectral width, and have good output power. They can outperform 1300-nm LED at a much lower cost in some applications.

Spectral Width

What is the spread of wavelength in the light? An LED emits a range of light, while a laser emits a single wavelength. Different wavelengths travel at different speeds through a fiber, which contributes to dispersion and thus limits the fiber's bandwidth. LEDs have spectral widths of around 25 to 40 nm. For an LED with a nominal wavelength of 850 nm, a 40-nm spectral width means that the wavelengths actually range from 830 nm to 870 nm. Figure 3-13 compares the typical spectral width of an LED and laser.

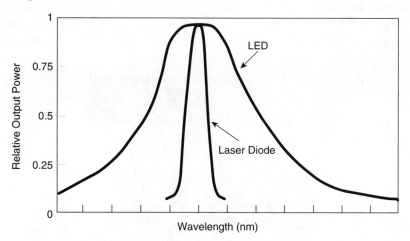

Figure 3-13. Source spectral width *(Courtesy of AMP Incorporated)*

Speed

How fast can the device be turned on and off? The speed at which the source can be modulated determines the data rate it can support. A laser can be turned on and off faster than an LED, so lasers are preferred for very-high-speed applications.

DETECTORS

At the other end of the link is the detector, which serves to convert the optical signal back to an electrical signal. The detector must be able to accept highly attenuated power – down in the nanowatt levels. Subsequent stages of the receiver amplify and reshape the electrical signal back into its original shape. Some detector packages have a preamplifier built in to boost the signal immediately.

Detectors operate over a wide range of wavelengths and at speeds that are usually faster than the LEDs and lasers. The important figure of merit for a detector is its sensitivity: what is the weakest optical power it can convert without error? The signal received must be greater than the noise level of the detector. Any detector has a small bit of fluctuating current running through it; this minuscule current is noise – spurious, unwanted current. The minimum power received by the detector must still be enough to ensure that the detector can clearly distinguish between the noise and the underlying noise. This is expressed as a signal-to-noise ratio or a bit error rate. SNR is a straightforward comparison of the signal level and the noise level, while bit error rate is a more statistical approach to determining the probability of noise causing a bit to be lost or misinterpreted.

TRANSMITTERS, RECEIVERS, AND TRANSCEIVERS

Packaged transmitters and receivers offer a convenient, cost-effective method of achieving the electro-optic interface. Typically in small, board-mounted packages, transmitters and receivers include the necessary electronics to permit building of a fiber-optic data link. Figure 3-14 shows 125-Mbps and 155-Mbps transceivers used for high-speed network applications.

Figure 3-14. Fiber-optic transceivers *(Courtesy of AMP Incorporated)*

A transmitter accepts standard TTL or ECL data, which it converts to light for transmission through the fiber. The receiver accepts the light and converts it back to TTL or ECL signals.

Today's packages are easier to use for a number of reasons. First, package dimensions and pinouts have been standardized so that packages from different vendors can be used interchangeably. Second, specifications for using the devices make building links easier. For example, the output power of a transmitter is now given as the power launched into a given type of fiber. Not too long ago, manufacturers listed only the total output power, plus some specs like NA and diameter. The designer was left to calculate and verify the actual power launched into the fiber he was using. Third, transmitter and receiver packages are offered in matched pairs. That is, you can buy an FDDI set, an ATM set, or an Ethernet set and know that they will meet the published specifications for achieving the speeds and distances allowed by the standards.

LINK POWER MARGIN

The difference between the power the source launches into the fiber and the minimum sensitivity of the detector defines the link power margin. Suppose, for instance, that the source launches -19 dBm into the fiber and the receiver sensitivity is -33 dBm. The link power margin is 14 dB. During transmission from end to end, from transmitter to receiver, the signal must not lose more than 14 dB of optical power. Sources of loss include not only fiber attenuation, but losses associated with interconnections at patch panels, wall outlets, and so forth. In addition, a prudent power margin reserves 3 dB for aging of the source.

Most networking and building cabling standards define a power margin. These margins are routinely achievable within the physical limits allowed by the standards. That is, the standards recommend cable distances and the number of interconnections along the path. By following these guidelines, the link should be well within the power margin. Fiber-optic networks typically have a link margin of about 11 dB.

A convenient way to look at a link budget is by graphing the losses along the path. Figure 3-15 shows a simple power margin graph. Fiber attenuation can be estimated by assuming loss is linear per unit length. Thus, if the attenuation is 1.5 dB/km, then the loss is 0.75 dB for a 500-meter run or 0.15 for a 100-meter span. Standards allow 0.75 dB loss for each interconnection. Allow 3 dB for aging. A quick addition of sources of loss in the link will allow a budget to be quickly calculated. In most cases, a measurement of link losses will show losses that are lower than this simple rule-of-thumb addition. One reason is that most interconnections provide well under 0.75 dB of loss, so that your quick addition is for worst-case conditions. If the measured loss is higher than the calculated loss, then there is probably something wrong with the installation.

Efficiency is nice; it is like getting as much optical power as possible from one end of the link to the other. Along the way, of course, there will be interconnections. Each interconnection will contribute some loss. The role of the connector or splice is to provide a mechanical coupling mechanism that minimizes loss.

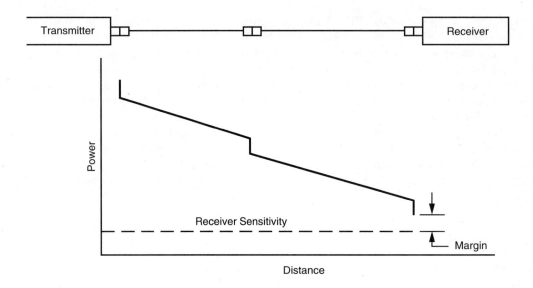

Figure 3-15. Power margin graph *(Courtesy of AMP Incorporated)*

Alignment: The Key to Low Loss

Interconnection is a critical part of a fiber-optic link. The job of a connector or splice is to provide low-loss coupling of light through a juncture. Low loss is the result of precise alignment of the two fibers being joined. Any misalignment of the two fiber cores increases the loss.

Losses stem from three sources:

1. Tolerances in the connector. As with any device, perfection is in the control of tolerances. A fiber-optic connector has tolerances associated with it that tend to allow some small amount of misalignment.

2. Tolerances in the fiber. The fiber should have a perfectly round core centered exactly within a perfectly round cladding. As these depart from perfection, loss is increased.

3. Different fiber types. Loss is increased if the receiving fiber has a smaller core or NA. In most premises cabling application, the same type of fiber is used throughout the installation, so these losses are not a concern. Such mismatches can contribute considerable loss. For example, coupling light from a 62.5/125-µm into a 50/125-µm can add up to 6 dB additional loss from the mismatches in core diameter and NA.

If tight connector and fiber tolerances are critical to low-loss interconnections, the good news is that both connectors and fiber are made with a high degree of precision. Today's connectors offer an insertion loss of around 0.3 dB – well under the 0.75 dB permitted by network standards. This means that connector performance far exceeds applications requirements.

FIBER VERSUS COPPER: THE DEBATE

If fiber is so great, why hasn't it replaced copper faster? Good question.

Pro-copper arguments run like this:

- Copper-based systems, especially UTP, are less expensive to buy and to install. This issue centers on the cost of the electronics, not the cost of the cable and related hardware. The electronics for fiber-based systems is more expensive than electronics for copper-based systems. The gap is narrowing, but copper retains an edge.

- UTP cables are well understood, with years of proven experience. Both craftspeople and users are comfortable with the technology.

- UTP has proven capable of meeting the evolving demands of networks, and there is no reason to believe that it cannot continue to do so.

- Fiber is a niche product. It's best applied in building backbones, in long spans between buildings, and in other situations that clearly require fiber.

- Fiber to the desktop is overkill – too costly and with more speed than most users need. Who needs 100 Mbps to the desktop? And even if you do, you can achieve it with low-cost UTP.

- Fiber is tricky stuff to work with.

Pro-fiber arguments run like this:

- Fiber is clearly superior technologically. Installing fiber ensures performance as even higher speed networks emerge in the future. Fiber optics "future-proofs" your building.

- There is no assurance that Category 5 UTP can continue to meet emerging needs. The last thing we need is Category 6 cable that makes today's premises cabling plant obsolete.

- Fiber-based systems are only slightly more expensive to buy and install. But because they are more reliable, the lifetime costs of the cabling system may prove to be lower than with UTP. For example, the costs associated with network interruptions, downtime, and repairs will be much lower because there will be fewer disruptions. Besides, prices for fiber-optic components are steadily declining and should continue to do so.

- Fiber to the desktop is not overkill. The history of computers and networking demonstrate that demand for higher performance is insatiable. Fiber is the better bet to "future-proof" your building.

- Fiber is no trickier than Category 5 cable to install.

Who's right? Both sides, of course, have valid and not so valid points when they push their arguments to extremes. For example, fiber hardware, such as a wall outlet, was once significantly more expensive than a copper-based outlet. Today some vendors price both copper and Category 5 hardware the same or nearly the same. Arguments over economics are slippery because they can change rapidly. What is clear is that at Ethernet and Token Ring speeds of under 16 Mbps, copper has a price advantage, although fiber is finding wider use in backbone and special applications. In high-speed systems of several hundred megabits per second or more, fiber is the hands-down winner. Long-distance telecommunications and high-speed computer interconnections prefer fiber today because copper just can't compete technologically or economically.

The murky area is around 100 Mbps.

CHAPTER 4

Connectors and Interconnection Hardware

Connectors and other interconnection hardware play an important role in premises cabling because they are the "glue" that holds everything together. Equally important, they provide the flexibility that allows the moves, adds, and changes that are part of the building's life.

The connector is applied to the end of the cable – terminating the cable, in the terminology of the industry. Improperly applied connectors can be a significant cause of problems in the cabling system.

TERMINATING A COPPER CABLE

The key to properly terminating a cable is to achieve an intimate gastight joint between the connector contacts and the cable conductor. Ideally, the connector will become a transparent extension of the cable. In reality, it does not. It increases the resistance a bit, can cause impedance discontinuities, can present a point of mechanical or electrical failure, and can add crosstalk. The gastight joint is important to ensure the long-term reliability of the connector by preventing moisture, gases, humidity, and other environmental effects from degrading the connection. Degradation comes, most often, in the form of increased resistance.

There are two main approaches to terminations applicable in premises cabling: crimping and insulation displacement (IDC).

With both crimping and IDC, it is important to use the correct size and type of wire and the correct tool. A contact, for example, might be designed for 22 to 24-gauge stranded conductors. Using a solid conductor or a 20 or 26-gauge conductor will result in a termination of dubious value – one more likely to fail.

Crimping

In the crimp, a portion of the contact is "crushed" under great pressure around the conductor. The conductor and contact fuse in a cold weld that provides the required electrical and mechanical properties. Crimping tools are designed to apply the proper pressure by closing the dies a fixed amount. An undercrimp results in high resistance and a loose connection. An overcrimp can damage the wire or contact so that they fail mechanically. Figure 4-1 shows the cross section of a crimp.

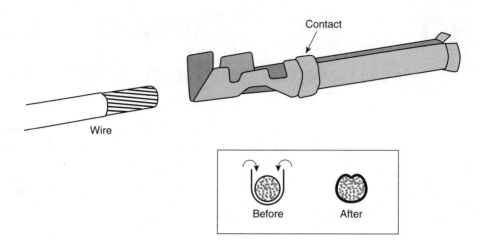

Figure 4-1. Crimp *(Courtesy of AMP Incorporated)*

Insulation Displacement

The insulation displacement contact uses a slotted beam. A wire is driven between the slot, the beams piece the insulation and deform the conductor. As with the crimp, an intimate gastight joint is formed. The beams maintain a residual spring pressure on the conductor to achieve long-term reliability. One advantage of IDC terminations is that the wire insulation does not have to be removed.

An IDC contact can either be a flat formed plate or a barrel. Some barrel-shaped IDC contacts allow more than one wire at a time to be terminated. IDC terminations are the most common in premises cabling applications.

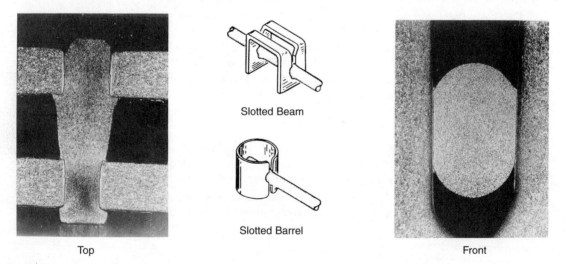

Figure 4.2 IDC termination *(Courtesy of AMP Incorporated)*

MODULAR JACKS AND PLUGS

Modular jacks and plugs (Figure 4-3) are familiar to everyone as connectors that plug into telephone handsets, bases, and wall outlets. While residential telephones use four-position plugs and jacks, UTP uses an eight-position connector to terminate all four pairs of the cable. In premises wiring, the plug goes on the wire and the jack goes on equipment, into outlets, and so forth. In the jacks in Figure 4-3, you can see both the barrel-style and 110-style contacts for terminating cables.

Figure 4-3. Modular plugs and jacks *(Courtesy of AMP Incorporated)*

Modular plugs and jacks are often referred to as *RJ* connectors from USOC specifications defining their use. An RJ-11 is a 6-position connector; an RJ-45 is an 8-position connector. As commonly used to indicate simply connectors, both terms are wrong. RJ11 and RJ45 refer to specific connectors, wiring patterns, and applications within the telephone system. To the telephone person, they have narrow, precise meanings. Network people have appropriated the terms to simply distinguish between 6- and 8-position connectors, however incorrectly, and these terms have gained wide acceptance.

Modular plugs use IDC terminations

In using modular plugs and jacks, here are some important things to remember:

- For Category 5 applications, you must have plugs and jacks rated to Category 5.

- Plug contacts sometimes differ depending on whether they are used with stranded or solid conductors. Some connectors can terminate both stranded and solid conductors; others can only be used with one type or the other. Do not use the wrong type of plug connector for the cable you are using.

- Follow the termination procedure carefully. With Category 5 cable in particular, following the proper installation procedures is essential to meeting 100-MHz performance.

UTP Connector Performance Requirements

As with cable, connectors are also subject to performance requirements to ensure their suitability to the applications. Originally, performance requirements were covered in EIA TSB-40, but this document became obsolete and the requirements were incorporated in TIA/EIA-568A. Figures 4-4 and 4-5 show attenuation and NEXT loss requirements for connecting hardware.

Modular jacks and plugs are also available in shielded versions for use with sUTP cable.

Frequency	Attenuation (dB)		
	Category 3	Category 4	Category 5
1	0.4	0.1	0.1
4	0.4	0.1	0.1
8	0.4	0.1	0.1
10	0.4	0.1	0.1
16	0.4	0.2	0.2
20	–	0.2	0.2
25	–	–	0.2
31.25	–	–	0.2
62.5	–	–	0.3
100	–	–	0.4

Figure 4-4. Attenuation for UTP connecting hardware

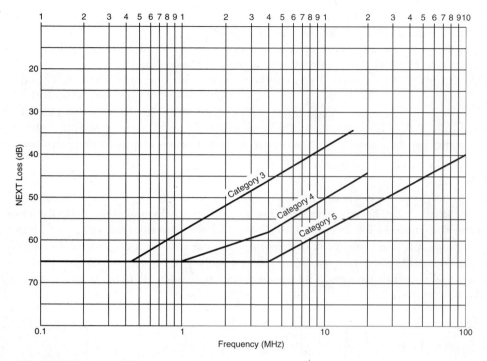

Figure 4-5. NEXT loss for UTP connecting hardware

Pinouts

Connector pinouts refer to the assignment given to each contact in the connector. In other words, which conductor of the UTP goes into which contact of the cable. Actually, there are three standard pinout configurations for UTP and modular jacks, as shown in Figure 4-6.

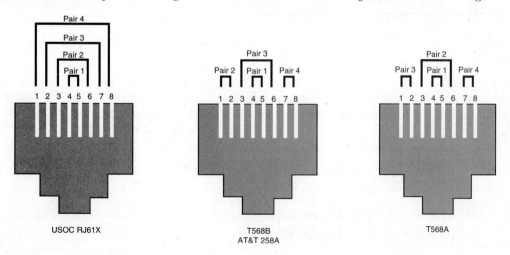

Figure 4-6. Modular plug and jack pinouts

USOC RJ61X. This configuration allows a 6-position plug to be inserted into an 8-position jack and still make the connection on Pairs 1, 2, and 3.

AT&T 258A / TIA/EIA-568B. Originally devised by AT&T, this configuration is also recognized by TIA/EIA-568 as an optional wiring pattern (hence the B).

TIA/EIA-568A. This standard represents a compromise between the USOC and AT&T wiring patterns and standards. It is backwards compatible with equipment of both the USOC and AT&T style. It is the preferred pattern in TIA/EIA-568A. Notice that the TIA/EIA-568A pattern and the AT&T pattern are identical except for the assignment of pairs. In other words, you can convert from a T568A to a T568B wiring pattern merely by relabeling the contacts.

Keying

Keying refers to methods to ensure that only certain plugs can mate with certain jacks. Some plugs have a key on one side or the other; mating jacks must have a similar cutout to accept the key. Figure 4-7 shows standard keyed and unkeyed interfaces of jacks. Notice that unkeyed plugs will also mate with keyed plugs. Keying can be used in premises cabling to clearly differentiate different types of lines. The most common use of keying is in the Digital Equipment DEC connect system. This system uses key jacks for data ports and unkeyed jacks for telephone ports.

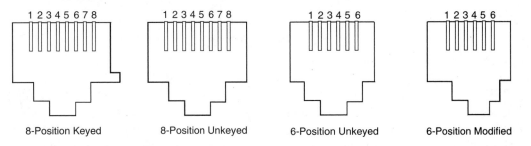

| 8-Position Keyed | 8-Position Unkeyed | 6-Position Unkeyed | 6-Position Modified |

Figure 4-7. Keyed and unkeyed modular jack interfaces

Wire All 8 Positions

Many applications use only two pairs (four wires). But different applications use different connector positions. To achieve a performance-driven cabling system that can be used with any application, it becomes important to terminate all four pairs. For example, 10BASE-T, Token Ring, and TP-PMD (for UTP on FDDI) all use different contacts to terminate cable pairs:

10BASE-T:	Pins 1-2 and 3-6
Token Ring:	Pins 4-5 and 3-6
TP-PMD:	Pins 1-2 and 7-8
ATM:	Pins 1-2 and 7-8
100BASE-T4:	Pins 1-2, 3-6, 4-5, 7-8

If you only wire an outlet with Pins 1, 2, 3, and 6 for Ethernet, the outlet can't later be used for Token Ring. So connect all four pairs.

DATA CONNECTORS

Data connectors (Figure 4-8) are 4-position connectors used with Type 1 STP in Token Ring networks. They use a hermaphroditic interface – that is, both halves of the connector have the same mating face.

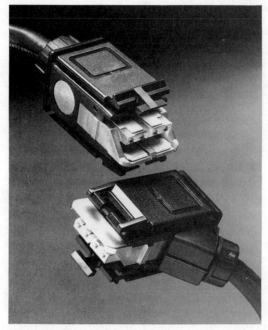

Figure 4-8. Data connectors *(Courtesy of AMP Incorporated)*

Data connectors short the transmitter and receiver pins when they are disconnected. This automatically wraps the signal through the connector to preserve the circuit when a cable is disconnected. The connectors use IDC barrel contacts and a stuffer cap to terminate pairs. The cable shield terminates in a metal ferrule. While data connectors are somewhat large and bulky compared to modular jacks, they are rugged and perform well. Although originally designed to operate at Token Ring speeds, newer versions are rated to 300 MHz.

25-PAIR CONNECTORS

Since a basic cable unit in telephone systems is the 25-pair cable, it's sometimes practical in premises cabling to run these cables as a unit. For example, a 25-pair cable can be connected directly from a PBX to a cross connect. Some network hubs also allow a 25-pair cable to run from the hub to the cross connect or patch panel.

Figure 4-9 shows the a 25-pair cable assembly. The connector is a miniature ribbon connector. The connector uses slotted-beam style IDC contacts to terminate the pairs; the mating face has a tongue-and-groove style interface. The "ribbon" part of the name derives from the shape of the contacts.

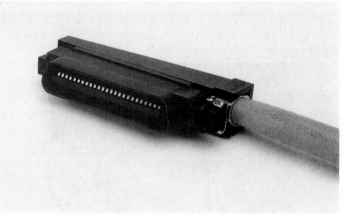

Figure 4-9. 25-pair cable assembly *(Courtesy of AT&T)*

Miniature ribbon connectors were originally designed for low-speed voice communications. While they are quite capable of being used in lower-speed applications such as 10-Mbps Ethernet, recent versions are rated to Category 5. Cross connects and patch panels are available that allow 25-pair connectors to plug in directly.

FIBER-OPTIC CONNECTORS

Fiber-optic connectors differ significantly from electrical connectors. Connectors for both copper and fiber have a similar function: to carry signal power across a junction with as little power loss and distortion as possible. But the difference between electrical signals and optical signals is that the connectors are very different. Where the electrical connector forms a gastight, low-resistance interface, the key to an optical connector is alignment. The light-carrying cores of the fibers must be very accurately aligned. Misalignment means that optical energy will be lost as light crosses the junction.

Alignment of the two fibers is a formidable task. Not only are the dimensions quite small – a core diameter of 62.5 μm for multimode fibers and 9 μm for single-mode fibers – tolerances complicate the matter. One hopes that the core is perfectly round and perfectly centered within a perfectly round cladding. And that the cladding is 125 μm, not 125.5 or 126 μm. The same applies with the connector dimensions. Unfortunately some variations creep into any manufacturing process. These variations represent misalignment that can increase the loss of optical power.

Even so, both fiber and connector manufacturers have evolved processes to create tightly toleranced components. Some years ago, the goal for insertion loss was 1.5 dB. Then 1 dB. Today's connectors offer loss of around 0.3 dB, while premises wiring standards call for a loss of 0.75 dB.

Loss is measured as insertion loss. Insertion loss is the loss of optical power contributed by adding a connector to a line. Consider a length of optical fiber. Light is launched into one end and the output power is measured at the other end. Next the fiber is cut in two, terminated with fiber-optic connectors, and the power out is again measured. The difference between the first and second measurement is the insertion loss.

Most connectors use a 2.5-mm ferrule as the fiber-alignment mechanism. A precision hole in the ferrule accurately positions the fiber. The two ferrules are brought together in a receptacle that precisely aligns the two ferrules through a tight friction fit.

Ferrules are typically made of ceramic, plastic, or stainless steel. Until recently, there were clear differences between the materials. Ceramic was more expensive and offered lower insertion loss. Plastic was the least expensive material, but offered higher insertion loss. Stainless steel fell in between. Today the price and performance of the materials is much closer, so there is little difference. The main exception is single-mode applications, where ceramic is still the preferred material.

Terminating a Fiber

Figure 4-10 shows the termination procedure for an epoxyless connector.

Epoxy has long been a staple for ensuring that the fiber is properly held in the ferrule. And epoxy has long proven that it does its job well. But epoxy has also been seen as a necessary evil. It has several drawbacks. It's an extra step in the process, it's potentially messy, and it requires curing. Curing time can be shortened with a portable curing oven, but that means another piece of equipment to lug around. Finally, when working in a completed building, the possibility of spilling epoxy on an executive's carpet or furniture is an unpleasant prospect.

Epoxyless connectors eliminate the need for epoxy and require only a crimp. Because early attempts at epoxyless connectors were less than successful, some people are reluctant to specify epoxyless connectors. But epoxyless connectors today offer low loss and long-term stability to make them a well-suited alternative to epoxy connectors.

Several approaches exist for eliminating epoxy. The most straightforward uses an internal insert that is forced snugly around the fiber during crimping. The insert clamps and positions the fiber, while the crimp secures the cable strength members.

The hot-melt connector uses preloaded adhesive so that external mixing is not required. The connector is placed in an oven for one minute to soften the adhesive. The fiber is then inserted into the connector; in three to four minutes the adhesive dries. The connector is then polished.

A third variation uses a short piece of fiber factory-assembled and polished into the end of the connector ferrule. The cable fiber butts against this internal fiber and the cable is crimped in place. This approach also eliminates the need to polish.

The main benefit of epoxyless connectors in premises cabling is increased productivity and, hence, lower costs.

> *For premises cabling applications, there is little performance difference between ceramic, plastic, and stainless steel-based connectors and between epoxy and epoxyless connectors. All can meet the requirements of premises cabling.*

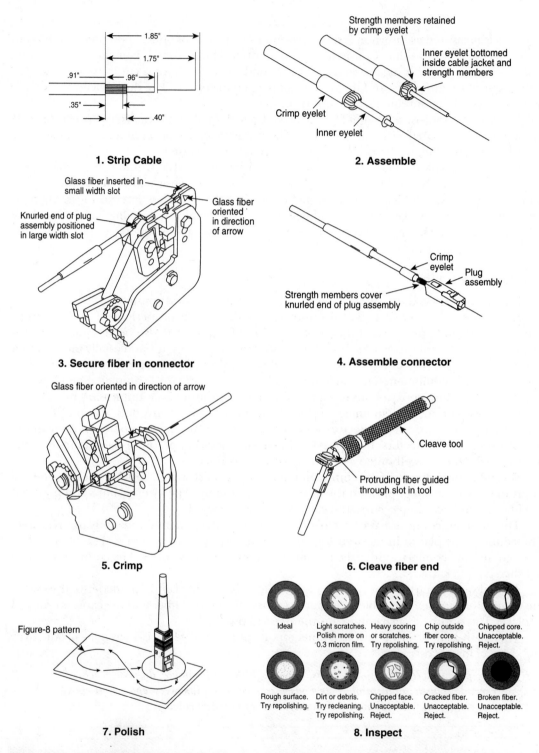

1. Strip Cable

2. Assemble

3. Secure fiber in connector

4. Assemble connector

5. Crimp

6. Cleave fiber end

7. Polish

8. Inspect

Figure 4-10. Terminating an optical fiber in an epoxyless connector *(Courtesy of AMP Incorporated)*

Types of Fiber-Optic Connectors

SC Connectors

Nippon Telephone and Telegraph (NTT) originally designed the SC connector for telephony applications. It has subsequently become the preferred general-purpose connector, and is being specified for premises cabling in TIA/EIA-568A, ATM, low-cost FDDI, and other emerging applications. Three characteristics make it popular.

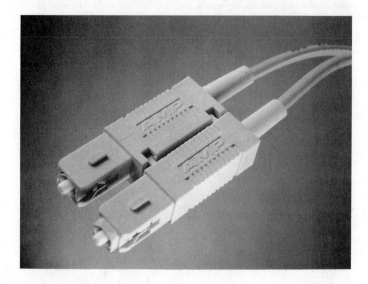

Figure 4-11. SC connectors *(Courtesy of AMP Incorporated)*

First, its design protects the ferrule from damage. The connector body surrounds the ferrule so that only a small part sticks out.

Second, it uses a push-pull engagement mechanism. Finger access is easier and requires less all-around space than a connector with a rotating engagement mechanism like threaded or bayonet-style coupling nuts. This permits closer spacings and higher density designs in patch panels, hubs, and so forth. And third, the rigid body eliminates intermittency problems caused by accidental pulling on the cable at the rear of the connector.

Connectors are easily snapped together to form multifiber connectors. Joining two connectors to form a duplex connector simplifies installation, use, and troubleshooting of the premises cabling plant. Since all links require two fibers, each carrying signals in opposite directions, a duplex cable assembly obviously makes sense. If you have 10 point-to-point links, it's easier to deal with 10 duplex cables rather than 20 individual cables.

TIA/EIA-568A recommends color-coding for SC connectors: beige for multimode connectors and blue for single-mode connectors.

ST Connectors

Invented by AT&T, ST connectors were the connector of choice during the late 1980s and early 90s, being replaced by the SC. The ST uses a bayonet coupling that requires only a quarter turn to engage or disengage the connector.

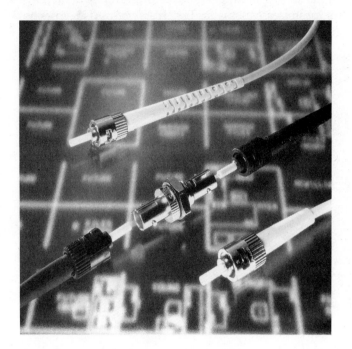

Figure 4-12. ST connectors *(Courtesy of AMP Incorporated)*

ST connectors are still an important force in premises cabling. Even TIA/EIA-568A acknowledges that they must be considered. Many buildings already wired with fiber use ST connectors. Most network equipment uses ST connectors for fiber-optic ports. So even new buildings will need to accommodate ST connectors in one way or the other in the future. One way is cables assemblies with STs on one end and SCs on the other. Another way is hybrid adapters that accept ST connectors on one side and SC connectors on the other side. Such hybrid adapters are available for nearly every combination of connector types.

FDDI Connectors

FDDI MIC (for medium interface connector) connectors are two-channel devices having two ferrules on 0.7-inch centers. The connectors use a fixed shroud to protect the fibers and a side-latch mechanism for keeping connectors engaged. While the connectors were designed for FDDI networks, they are not restricted to such uses. The connectors accept either factory- or field-installed keying to ensure that the cables are connected to the right FDDI station. Keying options are discussed in the next chapter.

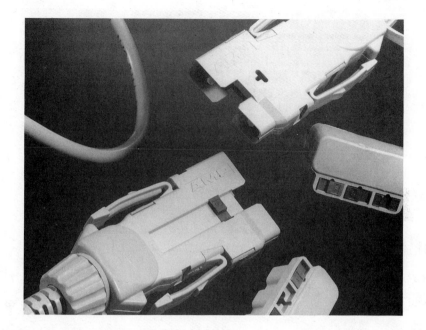

Figure 4-13. FDDI connectors *(Courtesy of AMP Incorporated)*

Splices

Splices are used to permanently or semipermanently join two fibers in cases where the need for connecting and disconnecting is not anticipated. We said "semipermanent" because some splices are re-enterable – that is, the splice is reusable.

The attraction of a splice is that it offers the lowest insertion loss – under 0.2 dB. Because compatibility is not an issue, splice designs tend to be proprietary to a single manufacturer. Fusion splices – which heat and actually fuse two fibers into a single unit – are used by telephone companies because they offer the lowest losses obtainable, as low as 0.05 dB. Fusion equipment is expensive – around $20,000 – so you need to perform thousands of splices before fusion techniques become cost-effective. For premises applications, mechanical splices that simply align two fibers precisely are more common.

Because disconnectable, rearrangeable circuits are central to a flexible premises cabling system, splices are not as common as connectors. They can be used for "behind the wall" splicing of long cable runs; typically, though, it is preferred to install long fiber lengths without splices. A second use of splices is for repairs of cable breaks.

Figure 4-14 shows a typical splice.

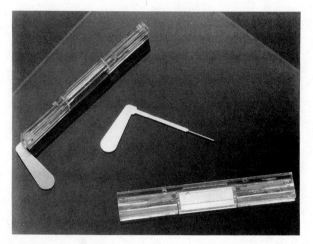

Figure 4-14. Fiber-optic splice *(Courtesy of AMP Incorporated)*

Coaxial Connectors

Coaxial connectors must terminate both the braid and the center conductor of the coaxial cable.

The most popular coaxial connector types are defined by military standard MIL-C-39012, which provides dimensions, materials, and performance requirements for a variety of connector types. From the military standards, many variations have evolved to provide lower costs, easier application, and greater flexibility.

BNC connectors are used with 10BASE-2 thinnet. They use a bayonet coupling and offer a 50-ohm nominal impedance, the same as the cable's. Figure 4-15 shows the BNC. BNC connectors are available in several styles:

- **Dual Crimp,** which is designed to meet rigid military specifications. The connector requires two crimps. First, the center contact is crimped to the center conductor. The connector is then assembled, and the braid is crimped to the connector body.

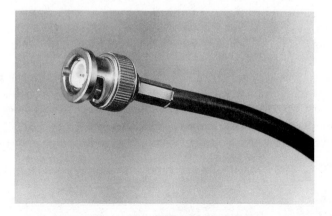

Figure 4-15. BNC connector *(Courtesy of AMP Incorporated)*

- **Single Crimp,** which requires only a single crimp cycle to terminate both the center and outer conductors. This significantly speeds the termination process.
- **Commercial,** which is a lighter, more compact, and less expensive single-crimp connector. It still meets the performance requirements of MIL-C-39012.
- **Consumer,** which is even less expensive than the commercial version. It does not meet all the performance requirements of MIL-C-39012; it does exceed the requirements of network applications.
- **Toolless,** which requires no special tools for termination.
- **Twist-on,** in which the termination is made by twisting a coupling nut onto the connector body.

Toolless and twist-on connectors are not recommended for network applications, except for temporary repairs. They are less able to withstand excessive flexing.

Coaxial connectors consist of a cable-mounted plug and a equipment-mounted jack.

Thinnet Tap

The Thinnet tap shown in Figure 4-16 is a popular connector for 10BASE-2 coaxial systems. The tap can fit into a standard electrical wall box or in many of the modular outlets used for premises cabling. The tap is installed with just a knife and screwdriver – no special tools are required. A companion drop cable plugs into the tap. One attractive feature of the tap is that it maintains a circuit regardless of whether the drop cable is plugged in. Moves, adds, and changes can be easily made with little disruption of the network.

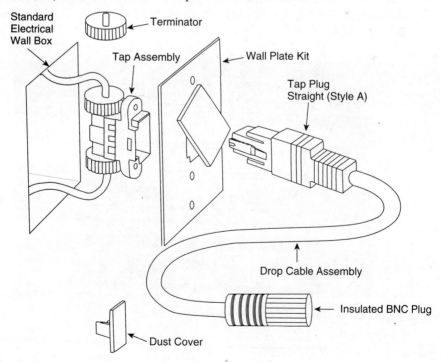

Figure 4-16. Thinnet tap for 10BASE-2 systems *(Courtesy of AMP Incorporated)*

Other Coaxial Connectors

10BASE-5 networks use an N-series connector. Larger than the BNC connectors, the N-connector uses a threaded coupling nut.

Video systems use a 75-ohm F-connector like those used by cable TV. The traditional F-connector does not have a center contact – it uses the center conductor of the cable as the contact. The original F-connectors were designed for analog signals at frequencies to around 216 MHz. Analog video signals can be very forgiving of poor performance from the connector. Consequently, F-connectors were designed more with cost in mind than performance. Traditional F-connectors are generally unsuited to digital data. Recently, however, more rigorous standards have upgraded the F-connector to make it suitable for digital signals and frequencies to 1 GHz.

Some mini- and mainframe computer systems use either twinaxial or triaxial cable. Connectors for these are similar to N-series connectors, but are often larger, bulkier, and harder to use.

PATCH PANELS

Patch panels provide a method of arranging circuits. For example, an employee who changes offices might want to keep the same phone number. By rearranging circuits at the patch panel, you can have the number ring in any office you want. Similarly, networks are re-arranged to add, subtract, and change workstations. Again, people can move around but stay physically attached to the same network by changing the circuits. The patch panel is the place where circuits are connected and re-connected. Figure 4-17 shows a patch panel, while Figure 4-18 shows a typical application.

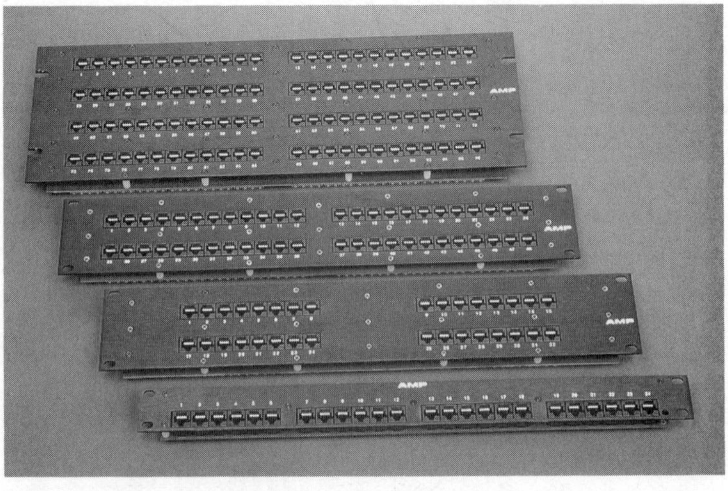

Figure 4-17. Patch panels *(Courtesy of AMP Incorporated)*

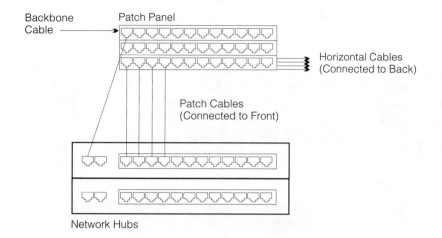

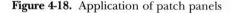

Figure 4-18. Application of patch panels

Many patch panels use a feed-through connector so that a cable can be plugged into both sides. For example, an RJ45 patch panel has a modular jack that accepts a modular plug in each side. The horizontal cables running to work areas plug into one side of the panel. Patch cables from equipment or from the backbone cable termination plug into the other side. While cables on either side of the panel can be rearranged, it is usually the patch cables that are changed. Think of the back side – the side with the backbone or horizontal cable – as being fixed and unchangeable in most cases. The front or patch side is where circuit reconfigurations takes place.

Feed-through patch panels are typically not suited to high-speed operation. Category 5 panels typically feature IDC contacts on the back and modular jacks on the front. The IDC contacts can be either 110 style or barrel style. This offers better electrical performance to reduce NEXT. Feedthrough couplers typically cannot meet Category 5 NEXT requirements.

Fiber-optic patch panels often provide a transition between different connectors. Transitions among SC, FDDI, and ST connectors are common.

PUNCH-DOWN BLOCKS

Punch-down blocks are so called because the wire is positioned in the top of an IDC contact and then forced (punched) down into the slot for termination by a tool called the punch-down tool. The tool also trims the wire by cutting off any excess on the end. Standard for telephone installations, punch-down blocks are the most common form of cross connect for twisted-pair cable. They're also called cross connect blocks. Like a patch panel, the purpose of the punch-down block is to provide the connection between two cable segments.

The two main varieties of punch-down cross connects are termed Type 66 and Type 110, which are AT&T designations.

Type 66 Cross Connects

The basic Type 66 block (Figure 4-19) has 50 rows of IDC contacts to accommodate the 50 conductors of 25-pair cable. Each row contains four contacts. The two left-hand contacts of each row are commoned; the two right-hand contacts are commoned. In use, one cable is terminated to the outer left-hand column. Another cable is terminated to the outer right-hand column. The two inner columns are used to perform the cross connect by terminating cross connect cables.

Type 66 blocks represent an older style designed originally for voice circuits. More recent versions meet Category 5 requirements. You should, however, be careful when using older blocks for digital circuits. Some older blocks have high crosstalk that makes them unsuitable even for moderate speeds.

Figure 4-19. Type 66 cross connect blocks *(Courtesy of The Siemon Company)*

110 Cross Connects

The basic 110 cross connect consists of three parts: mounting legs, wiring block, and connecting block. The mounting legs provide cable-routing management and the means for holding the wiring block. The wiring block consists of molded plastic blocks that position the cable. The block contains horizontal index strips. Conductors are laced into the slots of the index strip. The strip typically has 50 slots to accommodate a 25-pair cable. It is marked every five pairs to help visually simplify the installation and reduce errors. In addition, color coding helps identify the circuits. In application, the wires are laced into position and then punched into place with the impact tool. The wiring block, however, does not terminate the conductors. It simply positions them.

Figure 4-20 shows a 110 cross connect, while Figure 4-21 shows a closeup of the termination area.

Figure 4-20. 110 cross connect *(Courtesy of AT&T)*

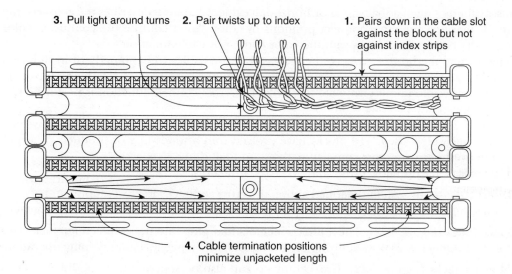

3. Pull tight around turns   **2.** Pair twists up to index   **1.** Pairs down in the cable slot against the block but not against index strips

4. Cable termination positions minimize unjacketed length

Figure 4-21. 110 cross connect termination area

After the wires are positioned in the wiring block, the connecting block is seated over the wiring block. The connecting block has IDC contacts at both ends. One set of contacts terminates the contacts in the wiring block; the other set is used for jumper wires that actually perform the cross connect. Once the connecting block is seated, the cross connect wires can be used to interconnect the cables properly.

Wiring blocks are available to accommodate up to 300 pairs. The basic structure is that each horizontal index strip handles 25 pairs; a cross connect uses multiple strips for higher pair numbers. A 100-pair cross connect requires four index strips; a 200-pair cross connect requires eight index strips. The connecting blocks come sized to accept either 3, 4, or 5 pairs. Each block has an alternating series of high and low teeth. Individual conductors of a pair are separated by low teeth; the high teeth separate pairs. The positions for each pair are color coded on the high teeth:

Pair 1:   Blue

Pair 2:   Orange

Pair 3:   Green

Pair 4:   Brown

Pair 5:   Slate

In addition, each contact for the pair is marked either "T" or "R," for tip and ring. Tip and ring are telephone terms that date back to the days of manual switchboards. The switchboard plug had three parts labeled tip-ring-sleeve. In modern multipair telephone cable, the ring conductor is color coded and the tip conductor is white.

Notice that the color coding of the connecting block matches the color coding of the pairs of a telephone cable. It does not refer to the function of the cable within the hierarchy of premises cabling – such as backbone or horizontal cable. Chapter 9 describes the color-coding system for marking the cross connect positions by color. Don't confuse the conductor color coding of the connecting blocks with the color coding by cable function.

The normal method of cross connection is that the cables to be joined are in different index strips. For example, the telephone backbone might terminate in row 1, the top strip. The horizontal cable might terminate in row 2. The cross connect function is performed by running a cable between rows 1 and 2.

We've just described the basic function and use of a 110 cross connect. As with anything as flexible and widely used, 110 blocks have evolved into a wider line of products. The two areas of interest here are cross connect methods and handling 25-pair cable.

110 blocks typically allow three types of cross connect:

- **Direct IDC:** In the system described above, the cross connect wires are simply run between connecting blocks. They are terminated by punching the conductors into the IDC contacts.

- **Patch Plugs** (Figure 4-22): These are connectors that plug into the connecting block. This eases rearrangements by simplifying the move, add, or change to a unplug/plug operation.

- **Modular Jacks** (Figure 4-22): Cross connects can also be supplied with modular jacks to perform the cross connect.

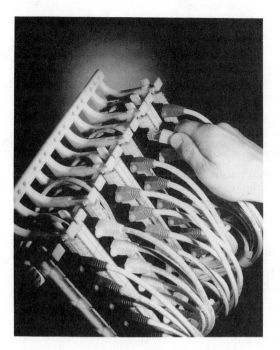
Modular Jack

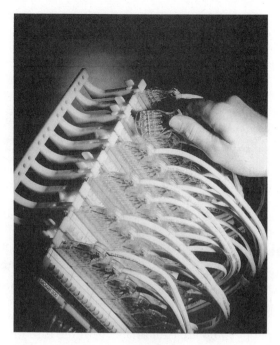
Patch Plug

Figure 4-22. Modular jack and patch plug 110 styles *(Courtesy of The Siemon Company)*

Cross connects can also be supplied prewired for specific applications. One popular variation uses a 25-pair connector. The block is prewired to the connector to allow a 25-pair cable from a device like a hub or PBX to simply plug into the cross connect without having to terminate individual pairs. Other versions terminate the backend cable in the same manner, but offer modular jacks for output.

The choice of cross connect style depends on many factors.

Cross connect blocks offer higher densities and so they require less space than patch panels. They are also less expensive. They can be wall mounted to reduce the amount of space required in a wiring closet, especially floor space, since they don't need a floor-standing rack. Larger systems are, however, rack mounted.

Direct IDC patching of cross connects is the least expensive and most straightforward of the various approaches. It is also the least friendly for making moves, adds, and changes. Skill is involved in removing and rearranging cables. Patch cables are easier, and modular jacks are easiest of all. Ease of use can be a two-edge sword. One edge is the ease and speed with which changes can be made. No skill is needed to unplug a modular plug from a jack and plug it into another jack. Skill *is* needed to reterminate patch wires. The other edge is security: do you want anybody to be able to perform moves, adds, and changes? Any user with access to a closet can make changes. Costs are also a consideration. In small systems, patch panels can make economic sense, but in larger installations the increase in cost may become unattractive. Fiber systems, of course, are patch panel-type systems.

APPLICATION GUIDELINES

Here are some suggestions for successful application of cross connect systems.

- Make sure the components are rated for the application.
- Group types of applications together.
- Use the cable management features of the block. Feed cables to be terminated on the right side into the right side of the block.
- Don't strip any more of the cable jacket away than is absolutely necessary.
- Maintain pair twist right up to the contacts. Maintaining twist becomes more important the higher the cable grade. With Category 5 cable, no more than 0.5 inch of a pair should be untwisted.
- Begin at the end that the cable enters. If the cable comes from the floor, start at the bottom and work upward. If it comes from the ceiling, start at the top and work down.
- Always double-check the routing of pairs, making any corrections immediately.
- Label all pairs. (Chapter 9 discusses documentation and labeling.)

The preference toward either cross connect blocks or patch panels may be part of your background. Cross connect blocks grew out of the telephone industry; those with telecommunications background are quite familiar with them. Patch panels are popular with network people, who are comfortable with the ease with which network connections can be made and rearranged. Telephone service has always been a company thing; the tradition of the telephone company or service provider maintaining the complete system, including any moves, adds, and changes, goes back many years. Even for companies committed to maintaining their own systems, the attitude of central control by trained people is strong.

Networks are moving in the other direction. The individualism of the PC carries over into networking. Many small networks were installed by users. Network administrators tend to want to have tighter and more immediate control over the network and its arrangement and favor patch panels. Patch panels are intuitively easy to understand and require no special training to use. Not so with cross connect blocks. Even as networks become a critical part of the company, with the same bureaucratic restraints, the legacy of patch panels remains.

The choice between cross connects and patch panels, of course, is not an either/or situation. Systems can also be mixed: a 110 block for voice, security, and other low-speed voice, and patch panels for high-speed data.

ELECTRONIC CROSS CONNECTS

Rearranging a cross connect or patch panel is basically a method of manually switching circuits on a permanent basis. In other words, the circuit was that way, now it's this way. Switching can also be done electronically. After all, telephone systems and some network devices operate by automatically switching circuits. The same technology can be applied to a cross connect.

The electronic cross connect (Figure 4-23) is typically a chassis that accepts plug-in modules that connect to a backplane in the rear of the chassis. The modules contain the ports into which cables are plugged. The unit allows any input port to be switched to connect to any output port. Switching is controlled through software. Figure 4-24 shows a screen from the control software.

Figure 4-23. Electronic cross connect *(Courtesy of NHC Communications)*

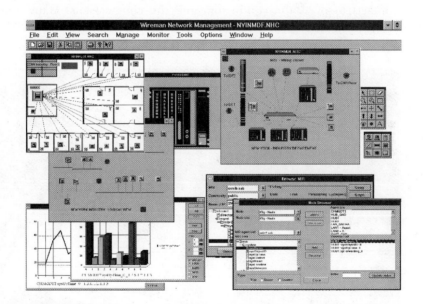

Figure 4-24. Software for electronic cross connect *(Courtesy of NHC Communications)*

Isn't electronic switching more expensive than manual switching? Yes, on the cost of equipment. But evaluating the benefits of electronic cross connects must include the costs of moves, adds, and changes. The ability to accomplish these quickly from a computer keyboard slashes the labor cost of changes. The software also permits better management of the system: screens can visually show the assignments of the ports. Some systems can also test the cable. As we will discuss further in Chapter 8, rearranged circuits should be tested to ensure that the link still meets specifications. The capability of rearranging and testing circuits automatically can significantly reduce the ongoing costs of routinely maintaining the cabling plant.

OUTLETS

Outlets are the transition between the behind-the-wall cable and the work area cables running directly to telephones, computers, and so forth. Again, there are many variations to meet different situations: wall outlets, in-floor flip-up outlets, surface-mount outlets for floors and walls, and outlets for modular office panels.

Modular outlets are increasing in popularity because of the flexibility they offer in creating and maintaining a cabling system. The idea of modularity is to provide a framework that will accept any type of cable and type of connector interface. Figure 4-25 shows one approach to a modular outlet that uses pluggable interfaces. Each interface – modular jack, BNC connector, data connector, and so forth – is placed on a small printed circuit board. This board, in turn, plugs into an edge connector in the outlet. The edge connector connects to the horizontal cable.

Figure 4-26 shows the basic setup of how the cable and modular insert meet in the edge connector. In the figure, the cable is held in a stuffer cap. This cap is pushed down over the IDC barrel contacts to press the conductors into the contact slots. Many newer modular approaches use a 110-style termination, especially for Category 5 applications.

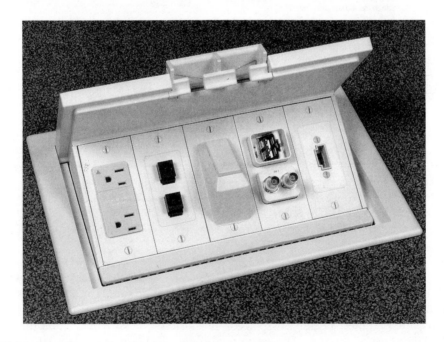

Figure 4-25. Modular floor outlet *(Courtesy of AMP Incorporated)*

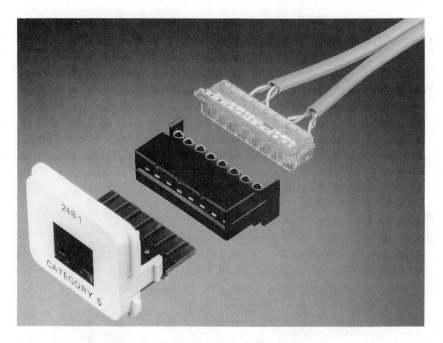

Figure 4-26. Close-up of structure of modular insert *(Courtesy of AMP Incorporated)*

What makes the system so flexible is that the interface modules can be changed without changing the horizontal cabling. The four-pair horizontal UTP, for example, always terminates in the edge connector in the same pattern. The wiring of the printed circuit board/modular jack insert can be changed to suit any need, general or application specific. Even though all edge connectors are wired identically, you can obtain modular jack inserts wired for 568A or 568B patterns, for Ethernet or Token Ring, for DEC connect or ISDN, or for telephone. You can plug in data connector interfaces for Token Ring or coaxial interfaces that include circuitry for converting from coaxial cable to UTP.

Modularity has several benefits:

- It allows both current, emerging, and legacy applications to be easily supported and even intermixed.

- It provides an easy upgrade path for technology without requiring excessive recabling or changes in hardware.

- It allows a single outlet to be flexibly wired for different applications. Many outlets accept four inserts. They can be mixed and matched.

- It promotes open systems by allowing both standard and proprietary inserts to be used to achieve vendor-independent cabling.

Other systems use a similar approach, but eliminate the edge connector. The plug-in interface (such as a modular jack) and the horizontal cable termination (such as a 110 block) are all part of a single module. Again, barrel and 110-style IDC termination predominate. Figure 4-27 shows several modular outlets.

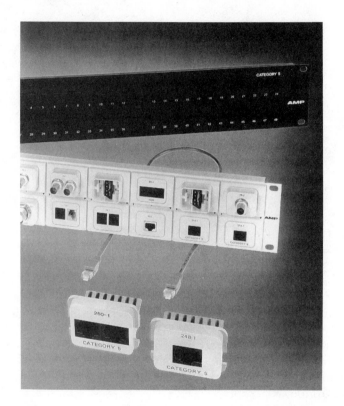

Figure 4-27. Patch panel using modular inserts *(Courtesy of AMP Incorporated)*

CHAPTER 5

Networks

Networks drive the performance of premises cabling. Category 5 cable, for example, results from the need to support higher data rates. Fiber has found renewed interest as data rates rise. While telephones are certainly an important part of premises cabling, their performance requirements are more modest than newer networks. Many applications can install Category 3 cable for telephones and Category 5 cable for networks, although installing Category 5 cable for all applications is growing in popularity.

As you saw in Chapter 1, an important feature of building cabling standards like TIA/EIA-568A and ISO/IEC 11801 is that they are application-independent. Rather their recommendations aim to achieve a level of performance that allows any application with the same performance requirements to operate on the cabling system. Ideally a 100-MHz cabling system will accommodate any network up to 100 MHz. Nevertheless, because the cabling system and the network are so intimately related in real life, this chapter provides an overview of local area networks. The perspective is the physical layer – that part that includes the cable and attached equipment.

LANS

A LAN is a sophisticated arrangement of hardware and software that allows stations to be interconnected and pass information between them. The *topology* of a LAN refers to its physical and logical arrangement. Figure 5-1 shows common network topologies.

A *bus* structure has the transmission medium as a central line from which each node is tapped. Messages flow in either direction on the bus. Most bus-structured LANs use coaxial cable as the bus.

A *ring* structure has each node connected serially in a closed loop. Messages flow from one node to the next in one direction around the ring.

A *star* topology has all nodes connected at a central point, through which all messages must pass.

A *tree* topology uses a branching structure. Most often this forms a *hierarchical star* layout such as is used in premises cabling. There are several star points, with each star center feeding a star higher up the chain. A single star occupies the top. Many practical configurations – including networks and premises cabling – use a tree topology.

Hybrid topologies combine one or more of the basic topologies.

It is important to distinguish between a *logical* topology and a *physical* topology. A logical topology defines how the software thinks the network is structured – how the network is "philosophically" constructed. A physical topology defines how the LAN is physically built and inter-

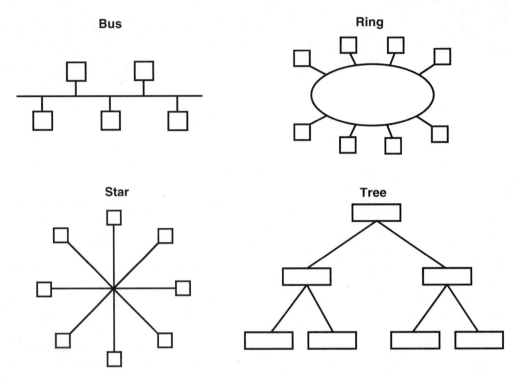

Figure 5-1. LAN topologies

connected. For example, the Ethernet LAN is a logically bus-structured network. Its physical topology depends on the specific variety of network. Sometimes it is wired like a traditional bus. Other times it is wired as a star or a hybrid configuration. Regardless of how it is wired or interconnected, the LAN appears to the control software as a bus network. Similarly, although Token Ring is a ring network, it is also configured as a star with all stations connecting through a hub.

Star-wired networks are popular because they increase reliability. Cable breaks in ring and bus structures can bring down the entire network. In a star network, only the affected portion crashes; the rest of the network can continue. The cabling topology recommended in TIA/EIA-568A and ISO/IEC 11801 is a hierarchical star configuration.

NETWORK LAYERS

As data communications becomes an important part of business, a need arises for universality in exchanging information within and between networks. In other words, there is a need for standardization – a structured approach to defining a network, its architecture, and the relationships between the functions.

In 1978, the International Standards Organization (ISO), issued a recommendation aimed at establishing greater conformity in the design of networks. The recommendation set forth the seven-layer model for a network architecture shown in Figure 5-2. This structure, known as the Open Systems Interconnection (OSI), provides a model for a common set of rules defining how parts of a network interact and exchange information. Each layer provides a specific set of services or functions to the overall network. The three lower layers involve data transmission and routing. The top three layers focus on user needs. The middle layer, transport, provides an interface between the lower three layers and the top three layers. Unfortunately, each layer has numerous standards that define the layer differently. The network systems in this chapter, for example, define the lower layers differently.

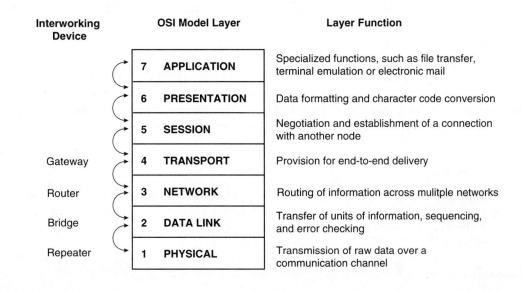

Figure 5-2. OSI network model

The following is a brief description of each layer.

Physical Layer

The physical layer is the most basic, concerned with getting data from one point to another. This layer includes the basic electrical and mechanical aspects of interfacing to the transmission medium. This includes the cable, connectors, transmitters, receivers, and signaling techniques. In other words, most of this book is concerned with the physical layer. Each specific type (or subtype) of network defines the physical layer differently.

While not used in LANs, the most widely used physical layer is RS-232, which defines the voltage levels of signals, the function of each wire, the type of connector, and so forth. The com or serial port of a personal computer implements an RS-232 physical layer. You can see that each specific need requires a different definition for this layer. Simply changing the cable type or connector changes the layer.

Data-Link Layer

The data-link layer provides for reliable transfer of data across the physical link. This layer establishes the protocols or rules for transferring data across the physical layer. It puts strings of characters together into messages according to specific rules and manages access to and from the physical link. It ensures the proper sequence of transmitted data.

Networks like Ethernet and Token Ring exist on the physical and data link layers. In a PC network, chips on the PC's network interface card perform the physical and data link functions. Software in the computer performs the functions of the higher layers. From the viewpoint of this book, building cabling and networks are distinguished at the physical and data link layers.

Network Layer

The network control layer addresses messages to determine their correct destination, routes messages, and controls the flow of messages between nodes. The network layer, in a sense, forms an interface between the PC and the network. It controls where the data on the network goes and how it is sent. The Internet uses protocols called transmission control protocol and internet protocols (TCP/IP) to control communications between computers on the net. IP works on the network layer.

Transport Layer

This layer provides end-to-end control once the transmission path is established. It allows exchange of information independent of which systems are communicating or their location in the network. This layer performs quality control by checking that the data is in the right format and order. TCP exists on the transport layer.

Session Layer

The session control layer controls system-dependent aspects of communication between specific nodes. It allows two applications to communicate across the network.

Presentation Layer

At the presentation layer, the effects of the layer begin to be apparent to the network user. This layer, for example, translates encoded data into a form for display on the computer screen or for printing. In other words, it formats data and converts characters. For example, most computers use the American Standard Code for Information Interchange (ASCII) format to represent characters. Some IBM equipment uses a different format, the Extended Binary Coded Decimal Information Code (EBCDIC). The presentation layer performs the translation between these two formats.

Microsoft Windows performs presentation layer functions.

Application Layer

At the top of the OSI model is the application layer, which provides services directly to the user. Examples include resource sharing of printer or storage devices, network management, and file transfers. Electronic mail is a common application-layer program. This is the layer directly controlled by the user.

In practice, n[...]p layer of one station (the message originator) to the top of anothe[...]ecipient). A message, such as electronic mail, is created in the top [...]e workstation. The message works its way down through the layer u[...]ed on the [...]nsmission medium by the physical layer. At the other end, the messa[...]ical layer and travels upward to the presentation level. It is at the prese[...]ctronic mail is read.

This layered appr[...]rk holds two benefits. First, an open and standardized systems perm[...]ent vendors to work together. Second, it simplifies network design, [...]g, enhancing, or modifying a network. For example, while most LA[...]coaxial cable or twisted pairs, adding a fiber-optic point-to-point link i[...]ayer. Higher levels of the OSI model do not care how the physical lay[...]ed; they care only that the physical layer follows certain rules in int[...]s.

Bear in mind that the [...]model. While the ISO has created protocol standards for each level, [...]Different systems use different protocols. For example, Novell NetWar[...]LAN M[...]ager use different protocols for the different layers. IBM's System [...]A) and Digital Equipment's Network Architecture (DNA) have yet [...]a system have to use all seven layers; a system can combine two layers [...]ols.

Since Novell NetWare is th[...]rk operating system for PCs, here are the protocols used:

7. Application: NetWa[...]

6. Presentation: Networ[...]

5. Session: NetBIO[...]

4. Transport: Sequence[...]ge (SPX)

3. Network: Internet Packet Exchange (IPX)

2. Data Link: Open Data-Link Interface (ODI)

1. Physical: (Specific network type: Ethernet, Token Ring, etc.)

A network can be connected to other networks of the same or different type. Sometimes users purposefully break a large network into several smaller segments to make the network more efficient.

Figure 5-2 also shows the devices required to interconnect either different networks or segments of the same network. The simplest device is a repeater operating at the physical layer. The repeater simply takes an attenuated signal, amplifies it, retimes it, and sends it on its way. It is a "dumb" device in that it does not look at the message; it only regenerates the pulses.

A bridge operates at the next level, the data link level, to link different networks that use the same protocol. The bridge differs from a repeater in its built-in intelligence. It only passes from one segment to another messages that are intended for the second segment. It controls which traffic passes through it and which remains local to the originating segment. The bridge examines the message and determines its destination. Bridging can help improve network efficiency by reducing the number of stations that a message must pass through.

A router operates at the network-control level and can handle different protocols. Routers offer greater sophistication than a bridge. First, they can handle different protocols so you can hook networks using unlike protocols together. Second, they can add additional information to a frame to allow routing over larger networks such as an X.25 packet-switching network used for long-distance transmissions over the public telephone system. Third, they can choose the best route if several routes are available. Many routers also compress the data to make routing, especially over slower-speed long-distance lines, more economical. For example, a router with 4:1 compression can reduce an 8-megabyte file to a mere 2 megabytes for transmission.

Bridges and routers can be used locally or remotely. In a local application, a single bridge or router connects each LAN segment. In a remote application, typically over the public telephone network or private or leased lines, a bridge or router is required at each end.

A gateway works at higher levels, serving as an entry point to a local area network from a larger information resource such as a mainframe computer or a telephone network.

The ISO model is a handy framework for understanding the structure of a network. Some network models do not even have seven layers.

ACCESS METHOD

Access refers to the method by which a station gains control of the network to send messages. Three methods are carrier sense, multiple access with collision detection (CSMA/CD); token passing; and demand priority.

In CSMA/CD, each station has equal access to the network at any time (multiple access). Before seizing control and transmitting, a station first listens to the network to sense if another workstation is transmitting (carrier sense). If the station senses another message on the network, it does not gain access. It waits awhile and listens again for an idle network.

The possibility exists that two stations will listen and sense an idle network at the same time. Each will place its message on the network, where the messages will collide and become garbled. Therefore, collision detection is necessary. Once a collision is detected, the detecting station broadcasts a collision or jam signal to alert other stations that a collision has occurred. The stations will then wait a short, random period for the collision to clear and begin again.

In the token-passing network, a special message called a token is passed from node to node around the network. Only when it possesses the token is a node allowed to transmit.

In demand priority, a central device acts as a broker for stations requesting access to the network. A station wishing to transmit sends a signal to the hub. If the network is idle, the hub immediately gives the station permission to transmit. If more than one request is received, the hub uses a round-robin arbitration to acknowledge each request in turn.

One feature of demand priority is that it allows two or more levels priority. For example, time-sensitive messages like video can be given higher priority than other transfers.

FRAMES

The information on a network is organized in frames (also called packets). A frame includes not only the raw data, but a series of framing bytes necessary for transmission of the data. Figure 5-3 shows the frame formats for both the Ethernet, FDDI, and ATM data transmissions. Notice that ATM has a short, fixed length format of 53 bytes. We will discuss the significance of this later in this chapter when we describe ATM. Other frames can also be used. For example, besides the data frame, a token-ring network also uses a token frame for passing the token around the network.

Preamble	Start of Frame	Destination Address	Source Address	Length	Data	CRC
7 bytes	1 byte	6 bytes	6 bytes	2 bytes	46-1500 bytes	4 bytes

IEEE 802.3 Frame

Preamble	Start of Frame	Packet ID	Destination Address	Source Address	Data	CRC	End of Frame + Status
≥ 2 bytes	1 byte	1 byte	2 or 6 bytes	2 or 6 bytes	0-4486 bytes	4 bytes	≥ 2 bytes

FDDI Data Frame

Header	Data
5 bytes	48 bytes

ATM Cell

Figure 5-3. Frame formats: Ethernet, FDDI, and ATM

Here is a brief description of the elements of an FDDI data frame.

The *preamble* indicates the beginning of a transmission. The preamble is a series of alternating 1's and 0's – 1010101010 . . . – that allows the receiving stations to synchronize with the timing of the transmission.

The *start of frame* is a special signal pattern of 10101011 that signals the start of information.

The *packet ID* identifies the type of packet, such as data, token, and so forth.

The *destination address* is the address of one or more stations that are to receive a message. In a network, each station has a unique identifier known as its address.

The *source address* identifies the station initiating the transmission.

The *data* is the information, the point of the transmission.

The *CRC* or *cyclic redundancy check* is a mathematical method for checking for errors in the transmission. When the source sends the data, it builds the CRC number based on the patterns of the data. The receiving station also builds a CRC. If the receiver CRC matches the transmitted CRC, no errors have occurred. If they don't match, an error is assumed, the transmitter is informed, and the transmission is sent again.

The *end of frame* informs the receiving station that the transmission is over. It also contains the status of the transmission. The receiving station marks the status to acknowledge receipt of the packet.

SHARED VERSUS SWITCHED MEDIA

Early networks like Ethernet and Token Ring use a shared-media concept in passing messages. The message goes to every station on the network, with each station checking the address to see if the message is for it.

More recently, switched media has become popular. In a switched-media network, a central device such as a hub contains a switching network. The hub checks the address of the message and then sends the message only to the correct station.

Switched-media networks rely on a star-wired configuration, since there must be intelligence at the star's center to decode the address and perform the switching. Network topologies like rings and busses must be shared media since the intelligence for checking addresses lies in the stations. The message must go to each station.

Figure 5-4 shows the difference between shared and switched media networks. Switched networks make much more efficient use of network resources. The time a message spends on the network is significantly reduced since it passes directly from the originating station through the hub to the destination station. Switching increases network availability and can serve as an alternative to a higher speed network. A 10-Mbps switched Ethernet network (rather than a 100-Mbps Fast Ethernet) may serve as a cost-effective upgrade to a 10-Mbps shared-media Ethernet.

The key difference between shared- and switched-media networks lies in the bandwidth available to each user. In a shared-media Ethernet network, the 10-Mbps bandwidth is shared by the network and attached users. As the number of stations increases, the bandwidth per station drops. In a switched Ethernet network, each station has its own 10-Mbps link. The aggregate or total bandwidth is determined by the central switch, which has a bandwidth many times greater than 10-Mbps.

TYPES OF NETWORKS

IEEE 802.5 Token Ring

IEEE 802.5 specifies a token-passing ring network operating at 4 or 16-Mbps. It is often called the IBM token ring because the original design was pioneered by IBM. While the LAN uses a *logical* ring, it is physically configured as a star network. Workstations attach through the network through a hub, and hubs are connected by network cable called the backbone or trunk as shown in Figure 5-5.

Token Ring originally ran on Type 1 STP, but has also been adapted for UTP and a UTP.

IEEE 802.3 Ethernet

IEEE 802.3 is a CSMA/CD LAN commonly called Ethernet, by far the most popular type of network in use worldwide. Ethernet, however, was originally a LAN defined by Xerox, Digital

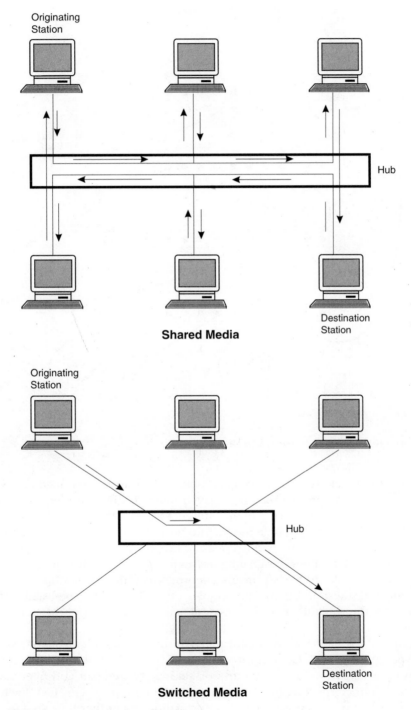

Figure 5-4. Shared versus switched media

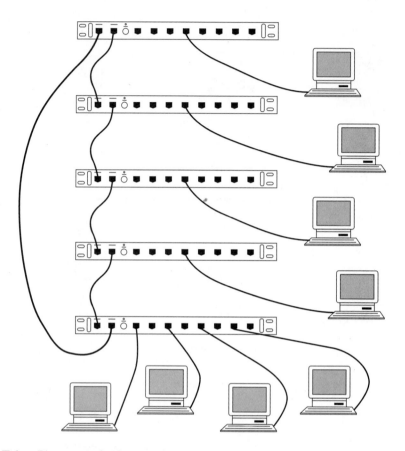

Figure 5-5. Token Ring network *(Courtesy of AMP Incorporated)*

Equipment, and Intel that used a thick coaxial cable as the transmission medium. Since then, the standard has evolved considerably to be much more flexible, although Ethernet is still widely used to describe this network. Minor differences exist between the Ethernet specification and IEEE 802.3. Still, most people use the two terms interchangeably, and so shall we.

Figure 5-6 lists several flavors of IEEE 802.3. The prefix number defines the operating speed, the middle word defines the signaling techniques, and the suffix defines the transmission medium. For example, 10BASE-T means a network operating at 10-Mbps, using baseband communication, and transmitting over twisted-pair cable. The cable type listed is the type of predominant cable for which the network was designed. Most flavors of Ethernet allow mixing of cable types.

While IEEE 802.3 is a logical bus topology, it is constructed as either a bus or a star depending on the type of cable used. Coaxial cables typically use a physical bus topology: the backbone cable runs serially from workstation to workstation. Twisted-pair and fiber use a logical star, with workstations connecting to a hub (also called a multiport repeater). Each hub contains several ports that permit attachment of a workstation or another hub.

Flavor	Cable Type	Segment Length* (meters)	Speed
10BASE-5	Thick coax	500	10 Mbps
10BASE-2	Thin coax	185	
10BASE-T	Cat. 3 UTP	100	
10BASE-FP	Fiber	1000	
10BASE-FB	Fiber	2000	
10BASE-FL	Fiber	2000	
100BASE-TX	Cat. 5 UTP	100	100Mbps
100BASE-T4	Cat. 3 UTP	100	
100BASE-FX	Fiber	2000	

* *Segment length is the backbone cable length for 10BASE-5 and -2. For others, it is the distance from hub to station (or other attached device).*

Figure 5-6. Flavors of Ethernet

10BASE-5

10BASE-5 – thick Ethernet – was the original version of the network, using a thick, rigid coaxial cable and N-connectors, as shown in Figure 5-7. It permits up to 100 stations over 500 meters for each segment. Up to three segments can be connected by repeaters, for a maximum of 1500 meters. 10BASE-5 is seldom found in new applications.

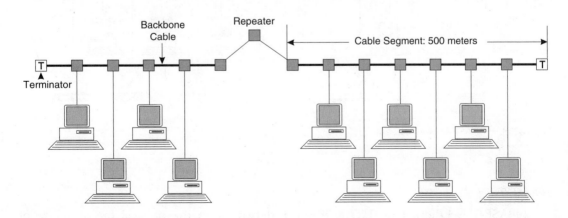

Figure 5-7. 10BASE-5 network

10BASE-2

10BASE-2 is thin Ethernet, using a flexible coaxial cable and BNC connector or thinnet taps. Each segment can be up to 185 meters long, with up to five segments per network. The distance from the backbone bus to the station is 50 meters maximum. Figure 5-8 shows a typical 10BASE-2 configuration.

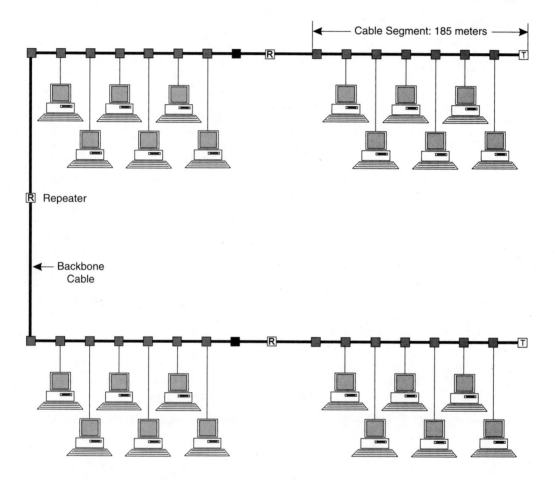

Figure 5-8. 10BASE-2 network

10BASE-T

10BASE-T is Ethernet on UTP. 10BASE-T departs from the physical bus structure of 10BASE-5 and -2 to use a hub-based star-wired configuration as shown in Figure 5-9. (Hubs have also been made with coaxial ports to make a star-wired 10BASE-2 network.)

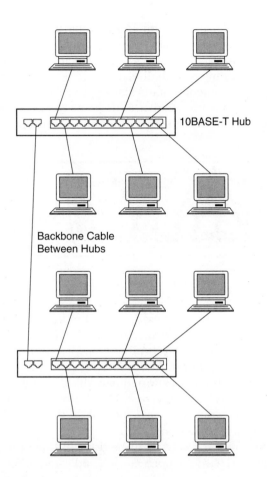

Figure 5-9. 10BASE-T networks

The hub is a repeater, so that each signal transmitted from any port on the hub has been amplified and retimed before transmission. 10BASE-T hubs are often called multiport repeaters.

10BASE-T networks can operate on Category 3, 4, or 5 cable. The maximum recommended distance from hub to station is 100 meters. While distances of 150 meters can be achieved with Category 5 cable, exceeding the 100-meter recommendation is not advised. Since the cable distance is from station to hub, two stations can be as far apart as 200 meters when they are connected to the same hub. Hubs, too, can be interconnected to extend the distance between two stations. Again, the recommended distance between two hubs connected over UTP is 100 meters. Many 10BASE-T hubs also allow fiber backbones that extend the distance between hubs to 2 km. You can see that the 100-meter station-to-hub limit is really flexible and not as limiting as it first appears.

A standard port acts as a transceiver for the type of cable being used; one function of the transceiver is to ensure that the signal is converted to the form required by the cable.

For example, a UTP uses two pairs, one for transmitting and one for receiving, and a modular-jack telephone connector. A coaxial cable uses only a single cable, allowing transmission in both directions over a single wire, and either BNC or N-type connectors. A fiber-optic cable requires two fibers, one for transmitting in each direction, and usually SC or ST connectors.

Another type of port is the attachment unit interface (AUI) shown in Figure 5-10. The AUI port is a 15-pin subminiature-D connector (similar to the 9- or 25-pin connector used in serial ports on personal computers) that does not include the transceiver function.

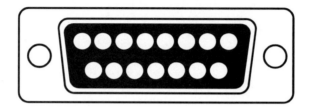

Figure 5-10. AUI port

An AUI allows different transceivers to be used. For example, consider a network adapter board in a workstation. You can buy a board specific to the type of cable you are using: a 10BASE-T board for twisted-pair cable or a 10BASE-2 board for thin coaxial cable. Alternatively, you can buy a board with an AUI port. The AUI port allows you to plug in an external transceiver for whatever cable you wish to use: twisted pair, thick or thin coaxial, or fiber optic.

10BASE-T networks are prone to cabling errors. Here's why. In the UTP cable, one pair carries signals in one direction; the other pair carries signals in the other direction. The transmitter at one end is connected to the receiver at the other end.

Connecting the transmitter to the receiver can be done in one of two ways as shown in Figure 5-11. The standard way is to have the ports at each end wired differently. Here the NIC card has a *standard* port: the transmitting pins are 1 and 2. The receiving pins are 3 and 6. The hub to which the NIC is connected is wired as a *crossover* port: the receiving pins are 1 and 2 and the transmitting pins are 3 and 6.

This means the cable runs straight through. Pins 1-2 on the transmitting NIC connect to wires 1-2 on the cable and pins 1-2 on the receiving hub. The transmitting lines on the NIC card are thus connected to the receiving lines on the hub.

Instances occur, however, when you want to connect two identical ports: say ports on two different hubs. If you use the same cable you used to connect the NIC to the hub, you are suddenly connecting the receiver to receiver and transmitter to transmitter. This doesn't work. Instead, a crossover cable is required. A crossover cable connects pins 1 and 2 at one end to pins 3 and 6 at the other end. This works. In most installations, crossovers will only be required in patch cables connecting hubs.

One hub can connect to another hub so long as IEEE 802.3 wiring rules are followed. The most important is the four-repeater rule. A signal passing from the originating station to the receiving station can pass through no more than four repeaters. Since a typical 10BASE-T network can easily have many hubs, care must be taken to ensure no signal paths violate the

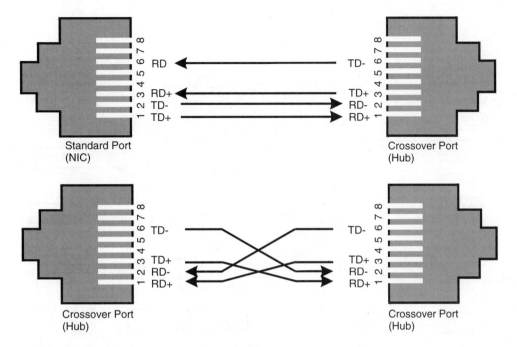

Figure 5-11. Standard and crossover ports and cables

rule. Figure 5-12 shows two methods of minimizing the number of repeaters in a path. One way is to use a single hub to feed multiple hubs. In this case numerous hubs can be used, but only three hubs are ever involved in the path. The second way, which has gained popularity in recent years, is stackable hubs. Stackable hubs are hubs designed so that when they are connected together, they appear to the network as a *single logical hub*. A signal can still pass through four logical hubs. The stackable hub shown in Figure 5-12 stacks four individual 12-port hubs to electrically connect them into a single logical hub having up to 48 ports. Each stack of hubs is treated as a single hub. Stackable hubs greatly simplify the task of avoiding wiring errors in interconnecting hubs.

10BASE-F

The 10BASE-F specification adds a fiber-optic alternative for use in IEEE 802.3 networks. The specification actually lists three different variations: a backbone cable (10BASE-FB), a passive star-coupled network (10BASE-FP), and a fiber-optic link between hub and station (10BASE-FL). Figure 5-13 shows a typical network containing all three variations; while Figure 5-14 is a table summarizing some of the principal optical characteristics of the 10BASE-F network.

The requirements for a 10BASE-FB and a 10BASE-FL are fairly straightforward. The backbone is simply a point-to-point connection between hubs. The FL versions define the connection between a hub and station or between a hub and passive star coupler. Notice that the hub is a multiport repeater that contains electronics to regenerate the signal. Each point-to-point link can be up to 2 km long.

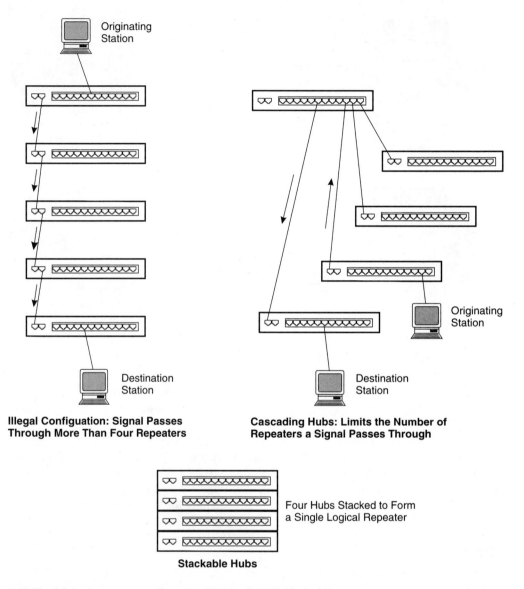

Figure 5-12. Ethernet repeater rules: cascading and stackable hubs

The 10BASE-FP option uses a passive fiber-optic star coupler to distribute light. The star can contain up to 33 ports. Ports can attach to a workstation or to a 10BASE-FL hub. The distance from the star to the attached station or hub is 500 m. Because the star coupler is passive, a 10BASE-FP network can connect workstations over an all-fiber network, with no electronics involved outside the workstation.

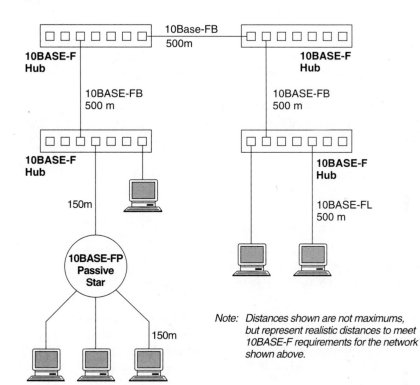

Figure 5-13. 10BASE-F network

	10-BASE-FP		10BASE-FB		10BASE-FL	
	Min	**Max**	**Min**	**Max**	**Min**	**Max**
Optical Wavelength (nm)	800	910	800	910	800	910
Spectral Width (nm)	–	75	–	75	–	75
Risetime (ns)	2	10	–	10	–	10
Launch Power (dBm)	-15	-11	-20	-12	-20	-12
Received Power (dBm)	-41	-27	-32.5	-12	-32.5	-12
Max. Length (m)	–	500	–	2000	–	2000
Connector Type	ST	–	ST	–	ST	–

Figure 5-14. 10BASE-F optical characteristics

IEEE 802.3 Fast Ethernet

Fast Ethernet brings a 100-Mbps upgrade path to Ethernet networks. Also known as 100BASE-X, Fast Ethernet uses the same frame format and medium access-control mechanism as its 10-Mbps sibling; it allows the same applications and network software, but at a tenfold increase in transmission speed.

Currently, Fast Ethernet recognizes three media types:

- 100BASE-TX uses Category 5 UTP and STP. The specification for this is derived from the TP-PMD originally created for FDDI on copper. Like 10BASE-T, 100BASE-TX does the crossover inside the hub.

    100BASE-TX, however, does not use the same RJ-45 pinouts as TP-PMD. To preserve compatibility with existing 10BASE-T wiring, it uses the same pinout. But, from TP-PMD, it uses MLT-3 signaling. For STP, it uses the same pinout as TP-PMD.

- 100BASE-FX defines a fiber-optic link segment.

- 100BASE-T4 uses four-pair cable to allow transmission over Category 3 cable. The standard recommends Category 5 connecting hardware for connectors, patch panels, punchdown blocks, and so forth. Of the four pairs, one transmits, one receives, and two are bidirectional data pairs.

100BASE-T defines two classes of hubs. Class I hubs can connect unlike media, so that segment types can be mixed. A Class I hub can have ports accepting two-pair UTP (100BASE-TX), four-pair UTP (100BASE-T4), and fiber. (Remember that, although 100BASE-TX only uses two pairs for signalling, it is usually wired with four-pair cable.) A Class II hub accepts only one media type.

Fast Ethernet allows 100-meter cable runs. In this case, the limit is set by timing considerations for the round-trip delay of a signal. It is built into the specification. In contrast, the 100-meter limit recommended for 10BASE-T is artificial in the sense that it is not a physical limitation of the network. With Category 5 cable, 10BASE-T can easily support 150-meter cable runs, although 100 meters is the generally accepted maximum.

IEEE 802.12 100VG-AnyLAN

100VG-AnyLAN, created by Hewlett-Packard and AT&T, is a switch-media network using demand-priority access methods and 100-Mbps operating speeds. 100VG-AnyLAN is very flexible, providing an upgrade path for both Ethernet and Token Ring (hence the name AnyLAN). More significant for building cabling, it will run on voice grade (Category 3) cable. How do you run 100 Mbps on a Category 3 cable?

100VG-AnyLAN uses a technique called quartet signaling, shown in Figure 5-15. The 100-Mbps signal is divided into four 25-Mbps signals. Because the network uses 5B/6B encoding, the transmission rate is 30 Mbps. Each of these 30-Mbps signals is transmitted on a different pair of four-pair cable. The data is also scrambled by the transmitter to ensure a good mix of 1s and 0s.

A similar technique is used with Token Rings running on 2-pair STP. The quartet signal is sent over the two pairs. 100VG-AnyLAN is also being adapted to use two-pair Category 5 cable and fiber-optic cable.

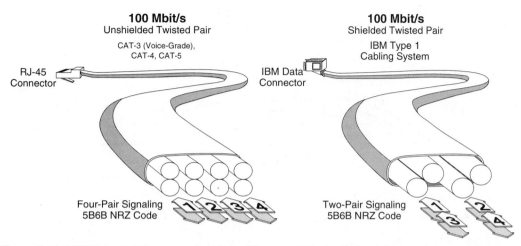

Figure 5-15. 100VG-AnyLAN quartet signaling *(Courtesy of Hewlett-Packard Company)*

One of the goals of 100VG-AnyLAN is to preserve the investment in building cabling. Buildings already wired with Category 3 or Type 1 STP cable do not have to upgrade. Only hubs and NIC cards need to be upgraded. (The catch here is those short-sighted building owners who did not install 4-pair Category 3 cable, but only 2-pair cable. In other buildings, 4-pair cable was installed, but only two pairs were terminated at the wall plate. Reterminating the cable is still much cheaper than running and terminating new cable).

FIBER DISTRIBUTED DATA INTERFACE (FDDI)

The Fiber Distributed Data Interface – FDDI – is the first local area network designed from the ground up to use fiber optics. Compared to its copper based counterparts, its performance is quite impressive: a 100-Mbps data rate over a 100-km distance and having up to 1000 attached stations.

While the FDDI standard recommends 62.5/125-μm fiber for multimode applications, it also allows 50/125-, 85/125-, and 100/140-μm fibers, as long as these do not exceed the power budget or distort the signal beyond the limits allowed by the specification. The power budget does not include optical loss at the interface between the source and connector. The specification calls for a minimum power to be launched into the fiber. This simplifies loss calculations. The power budget between stations must only include the fiber and any interconnections along the path.

FDDI is arranged as a token-passing ring topology that uses two counter-rotating rings as shown in Figure 5-16. The primary ring carries information around the ring in one direction, while the secondary ring carries information in the other direction. The reason for two rings is redundancy: if one ring fails, the other is still available. Redundancy lessens the likelihood of network failure. Further protection lies in the fact that each station is attached only to adjacent stations in the ring. If a station fails or a single point-to-point link fails, the network still functions. If a cable break occurs between two stations, a station can accept data on the primary link and transmit on the secondary link, as shown in Figure 5-17. This wraparound function makes FDDI highly reliable.

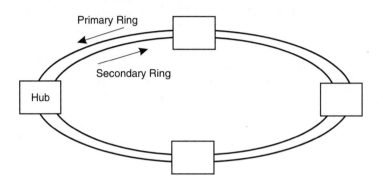

Figure 5-16. FDDI network

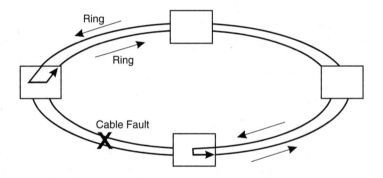

Figure 5-17. FDDI ring wrap from cable fault

When the secondary ring is not needed for redundant operation, both the primary and secondary rings can be used for data transmission to effectively double the network transmission speed to 200 Mbps.

Several variations of FDDI have evolved. From the cabling perspective, these are usually referred to as the *PMD,* for physical medium dependent. The PMD defines the requirements for the physical layer of the network. The main varieties are as follows:

Multimode Fiber PMD. The MMF-PMD is the original FDDI specification for multimode fiber. It uses the FDDI duplex MIC connector and provides keying for both the port type and polarity. It permits links lengths of 2 km between stations, with an 11 dB loss budget.

The preferred fiber is the well-known 62.5/125 fiber, although 50/125, 85/125, and 100/140 fibers are also allowed.

Low-Cost Fiber PMD. In an effort to lower the cost of FDDI links, LCF-PMD evolved. After looking at several alternatives in reducing costs – including using different fibers (such as 200/230 step-index fiber and plastic fiber), the FDDI committee decided the real cost of the link lay in the transceiver, not in the fiber. One approach – that of using less expensive 850-nm devices –was discarded because of the difficulties involved in ensuring that each end of the link was the same. The LCF-PMD relaxes the performance requirements for transmitters and receivers and replaces the FDDI connector with a lower cost SC connector. The resulting link length is 500 meters, only 25% of the distance allowed by MMF-PMD but still a considerable length for intrabuilding runs. The allowed link power budget is reduced from 11 dB to 7 dB.

For LCF-PMD, the 62.5/125 fiber is still preferred. Besides the same alternative fibers allowed by MMF-PMD, the LCF-PMD also allows 200-230 fiber. Unlike MMF-PMD, which keys connectors for both port type and polarity, LCF-PMD keys connectors only for polarity.

Single-Mode Fiber PMD. SMF-PMD covers single-mode fibers. The standard defines two categories of transceivers: Category I transceivers operate at the same power levels as MMF-PMD transceivers, while Category II transceivers are more powerful. The loss budget for a Category II transceiver is 32 dB. Transmission distances are about 40 km for a Category I link and 60 km for a Category II link.

Twisted-Pair PMD. TP-PMD allows the use of UTP and STP in FDDI links. It permits 100-meter runs between stations. Because other high-speed networks are adopting TP-PMD, we will discuss further in the next section.

Figure 5-18 gives a brief comparison of the different flavors of FDDI.

Characteristic	PMD			
	MMF-PMD	**LCF-PMD**	**SMF-PMD**	**TP-PMD**
Cable Type	Multimode Fiber	Multimode Fiber	Single-Mode Fiber	Category 5 UTP or Type 1 STP
Wavelength (nm), nominal	1310 nm	1310 nm	1310 nm	–
Link Distance (meters)	2000	500	60,000	100
Power Budget	11	7	32	–
Connector	Duplex MIC	SC	Duplex MIC	Modular plug or DB-9
Code	NRZI	NRZI	NRZI	MLT-3

Figure 5-18. FDDI PMDs

FDDI Stations

Each station contains either one or two transceivers. A dual attachment station (DAS) contains two transceivers for connection to both the primary and secondary ring. A single attachment station (SAS) has a single transceiver for connection with the primary ring. The SAS workstation does not attach directly to the network backbone, but through a hub. The hub may connect to both the primary and secondary ring of the fiber backbone, but only through a single connection to the workstation. This preserves the fault-tolerant redundancy on the backbone. If a SAS fails, the fault is isolated between the station and the hub.

Having a transceiver in each workstation eliminates the need for taps, repeaters, amplifiers, or other external signal conditioning equipment. Each cable is a point-to-point link to the next station.

Since each station or hub is a repeater, you might think that the network could be extended to an infinite number of nodes over an infinite area. After all, the signal is regenerated at each station. FDDI, however, places another limit on the network: the round-trip time of a message from the transmitting station. When a station transmits, the message goes from station to station. Each station reads the message, checking to see if it is the recipient and checking for errors. It then passes the message to the next station. When the destination station receives the message, it changes the status byte on the frame. The message then continues around the ring back to the transmitting station. The station verifies that the message has been received by checking the status byte and that no errors have occurred in the message's content. The maximum number of nodes and network length is limited by the time it takes this round trip. This time is adjustable in the SMT software to allow very large networks to exist.

FDDI also uses hubs to connect to both the primary and secondary backbone rings, but only form a single connection to SAS workstations. This simplifies the network layout and cuts costs without endangering the redundancy of a dual backbone.

Connector Keying

FDDI uses an optional keying system to prevent a cable from being connected to the wrong station port. Most connectors, both electrical and optical, are polarized to prevent mating upside down, but FDDI goes a step further to key each connector.

All MMF-PMD station ports are labeled as either A, B, S, or M according to the following:

A: port on dual-attach station; this port is the Primary In/Secondary Out port.

B: port on dual-attach station; this port is the Secondary In and Primary Out port.

M: master port on a hub to which a single-attach station connects.

S: A universal key. An S-keyed plug will plug into any port on the network, regardless of its keying. The S represents a single-attach station. S-keyed cables should be used with care since they defeat the purpose of keying – making sure the proper cable is plugged into the proper receptacle.

Notice that in a dual-attach station each port serves both the primary and secondary ring. Since an FDDI transceiver has both a transmitter and receiver, the receiver can accept signals on the primary in one direction, while the transmitter serves to transmit signals on the secondary in the other direction. Figure 5-19 also shows this idea graphically.

Cable connectors are keyed to match the keying of the receptacle connector on the hub or station. Cable connectors have two additional keying options:

AM: connects to either A or M port

BM: connects to either B or M port

There are three main types of cabling errors in FDDI:

1. Cabling for a dual-attach station is reversed. Its primary cable is connected to the secondary ring and vice versa. Keying one connector as A and the other as B prevents this error.

2. A single-attach station is connected on the trunk ring, causing a break in the backbone.

3. The M port of a hub is connected to the trunk ring, also causing a break in the backbone.

Single-mode fibers use a similar keying system, but with an "S" prefix: SS, SM, SA, and SB. Single-mode fibers have an additional key to allow only single-mode plugs and receptacles to mate. You cannot plug a multimode MIC connector into a single-mode receptacle. LCF-PMD does not use such keying.

Figure 5-19 shows the keying and some possible applications.

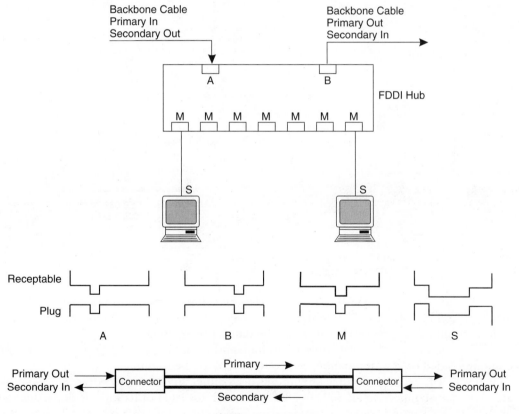

Figure 5-19. FDDI keying

FDDI Applications

An FDDI network is often used in conjunction with other networks. It can be used as a backend network, a frontend network, or a backbone network as shown in Figure 5-20.

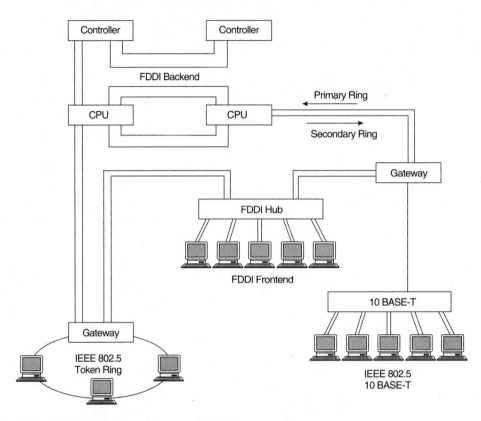

Figure 5-20. FDDI application *(Courtesy of AMP Incorporated)*

In a backend network, FDDI is used to connect mainframe computers, storage devices, controllers, and peripherals. Communication between these devices is one of the bottlenecks that slows computer operation. FDDI keeps the data zipping along.

As a frontend network, FDDI connects to high-end workstations used in such applications as computer-aided design, simulation and control, and high-end publishing.

Finally, FDDI can serve as a backbone network connecting other networks. For example, a large company could have an 802.3 or 802.5 network for each of its various departments – engineering, accounting, sales, manufacturing, marketing and so forth. Each individual network connects to the FDDI backbone through a gateway or bridge. FDDI then serves to connect all the different networks together. Not only is speed important here, but the capability of running up to 2 km between stations allows widely separated networks to communicate easily.

TP-PMD

TP-PMD (twisted-pair physical medium dependent) is an ANSI specification for 100-Mbps operation over UTP or STP. TP-PMD was originally devised for FDDI over copper. It has, however, since been adopted for 100-Mbps ATM and Fast Ethernet. TP-PMD uses 4B/5B, MLT-3 encoding to achieve a 100-Mbps data rate and 125-Mbps transmission rate. The information is also scrambled to spread the frequency content of the signal.

TP-PMD is intended for connections between hub and stations, not for backbone applications. For backbones, fiber remains the media of choice.

Recommended cables are Category 5 UTP and Type I STP. For UTP, the recommended connector is shielded or unshielded modular plug and jack. For STP, the recommended connectors are 9-position subminiature-D connectors.

ATM

Asynchronous transfer mode (ATM) is thought by many to be the network that will emerge the winner in the high-speed wars. What does ATM have going for it?

First, its basic operating speed is 155 Mbps. It can also work at 100 Mbps, using the same chipsets and signaling techniques as FDDI. For runs to the desktop, additional proposals are being considered for both 25 Mbps and 51 Mbps.

Equally important, ATM scales upward to higher speeds and is compatible with SONET. SONET is the fiber-optic scheme used by telephone companies for high-speed long-distance communications. SONET speeds start at 51 Mbps and scale upward in multiples of the basic speed to 10 Gbps and above. SONET does not have a top-end defined, but today's SONET systems top out around 10 Gbps. The attractiveness of this capability is that both LAN users and the telephone companies can adopt ATM. Because wide-area networking is becoming evermore important, ATM provides a straightforward way to use the telephone system as part of your network. Standards are being set to allow ATM to operate at 622 Mbps and 2.48 Gbps (both of which are SONET speeds). Such speeds, however, will require some reconsideration of building cabling. At such speeds, fiber will be the preferred medium, with single-mode fiber required for long backbone interconnections.

Second, ATM is a circuit switching network. ATM can establish a virtual circuit between two stations, allocating and guaranteeing the bandwidth required for the applications. This virtual circuit is very similar to a telephone call. The telephone company switches your call to the other end. As far as you and the other party are concerned, you have a dedicated circuit for as long as the phone call lasts. When you hang up, the circuit is broken. The difference between the voice call and the ATM virtual circuit is that ATM can guarantee exactly the amount of bandwidth required. As many users require varying amounts of bandwidth, ATM hubs must allocate the bandwidth dynamically. It can distinguish between time-critical and other requirements to prioritize communications between stations.

Third, ATM offers compatibility with existing Token Ring and Ethernet networks through emulation techniques. It is possible for existing low-speed networks and high-speed ATM to coexist. Like FDDI, ATM can serve as a high-speed backbone connecting other types of LANs.

53-Bytes

Token Ring, Ethernet, and FDDI use variable packet sizes. An Ethernet frame, for example, can range in length from 64 to 1518 bytes, while an FDDI frame can be from 20 to 4506 bytes long. One drawback to variable-length frames is that it becomes difficult to determine how long a frame takes to transmit on the network. In contrast, ATM uses a fixed length 53-byte cell. The cell is divided into two sections called the header and the payload. The 5-byte header carries addressing information; the 48-byte payload carries the information – voice, data, or video.

NETWORK DEVICES: HUBS

Many general and specialized devices are available for networks. With a star-wired network, the hub is the most important since it serves as the center of the star. From the perspective of premises cabling, hubs are fairly straightforward devices. They provide the ports into which cables plug to interconnect the network.

Hubs come in all sizes and levels of sophistication. At one end is the workgroup hub. While workgroup hubs are often associated with small groups, they can scale to accommodate many users. Their identifying feature isn't the number of users so much as their fixed functions. This is especially true with the growing popularity of switched and stackable hubs, which provide a ready growth path for expanding a network. Rather the workgroup hub is most characterized by its narrow breadth of functions. An Ethernet workgroup hub, for example, is just that: a hub designed for 10BASE-T networks. While many workgroup hubs have a certain modularity to allow them to be adapted to different needs, they still offer a more limited repertoire than an enterprise hub.

The enterprise hub occupies the high ground of network sophistication. These are chassis-based units that accept plug-in modules. A high-speed backplane connects the modules. The high gigabit speed of the backplane is important to keep the hub from becoming a bottleneck in network performance. The plug-in modules offer a wide range of functions: network station and backbone connections for any type of cable, bridges and routers, management functions to give network administrators greater control over monitoring and configuring the network, and so forth. Most enterprise hubs are also multiprotocol devices to allow several types of networks to be controlled from a single hub. For example, the hub can allow you to mix modules for 10BASE-T Ethernet, Token Ring, and FDDI. Each can exist independently or they can be connected by router modules.

Many networks will use a combination of hubs of different sizes and capabilities. For example, an enterprise hub can be located in the main cross connect closet. This hub serves as the "super center" in a expanding star network. This hub can then provide a backbone connection to a workgroup hub on each floor. The workgroup hub then serves users on that floor.

There is no clear separation between workgroup hubs and enterprise hubs. Small hubs are increasing in sophistication to take on capabilities once reserved for larger hubs. Other workgroup hubs are offering smaller size, much lower price, and attractive capabilities. An 8-port stackable hub can be as small as a VCR tape. At the same time, the traditional enterprise hub is appearing in more modest (and inexpensive) form to serve the needs of the workgroup. An enterprise hub typically has from 12 to 20 slots. A smaller version, with less than 8 slots, offers the same flexibility in configuring a hub, but on a scale more modest and better suited to a workgroup or small enterprise.

Figure 5-21 shows an enterprise-level multiprotocol hub.

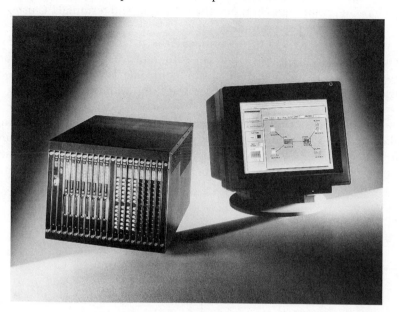

Figure 5-21. Multiprotocol hub *(Courtesy of Connectware)*

MINIS AND MAINFRAMES

A building cabling system must also accommodate the cabling requirements of minicomputers and mainframe computers. Before the days of open systems, LANs, and PCs, vendors of minicomputers and mainframe computers used proprietary methods to connect terminals to the computer system. While there are many variations, the two most important are the IBM 3270 system and the IBM AS/400 and S3x system. The discussion of how these systems are connected and how they can be integrated into a structured cabling system can easily be extended to other systems.

The IBM System Network Architecture is IBM's version of the seven-layer OSI model. It provides the layers of protocols to allow IBM equipment to operate together, including a scheme to allow the myriad of mainframe, minicomputer, and even PCs to connect.

At the heart of IBM's system is the 3270 family of products. Many, but not all, of the products begin with the digits "327." Here are some representative models:

3278 and 3279	Terminal
3279 and 3274	Terminal cluster controller
3299	Multiplexer
3245	Communications controller
3705 and 3725	Front-end processor

In a typical configuration, several terminals connect to a terminal cluster controller, which acts as a gathering point for messages between the computer and terminals. Cluster controllers are typically located locally to the terminals.

Groups of cluster controllers, in turn, connect either to a communications controller or front-end processor (FEP). These devices further gather, process, and communicate between the cluster controllers and mainframe. The cluster controller is often remotely located from the communications controller or FEP, communicating with them over the telephone network.

The reason for all this gathering and processing is that mainframe computers are very fast, while communications to the terminal are very slow. You don't want the mainframe wasting time doing "housekeeping" and waiting for slow communications. The communication controller and front-end processors form a bridge between the high-speed computer and the low-speed terminals and cluster controllers.

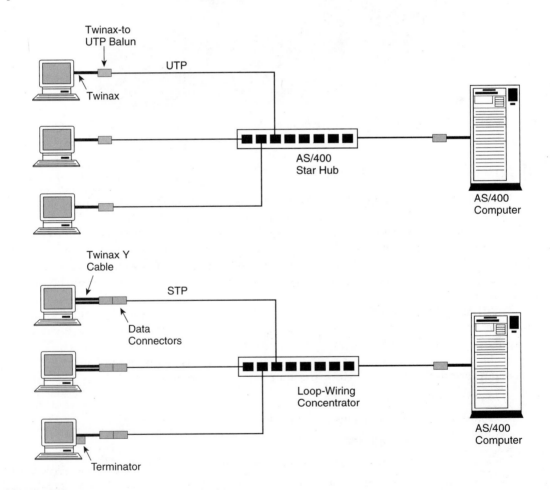

Figure 5-22. AS/400 system interconnections

AS/400 and Systems 34, 36, and 38 refer to IBM midrange computers. In the original configuration, up to seven devices can be daisy-chained together serially. Each daisy-chain connects to a communication channel on the host computer. Because the daisy-chain system is not easily adaptable to other cabling schemes, a star configuration was developed. How terminals connect to the host computer depends on the system used and the type of wiring. Figure 5-22 shows several examples of how AS/400 systems are interconnected.

STP: Up to seven terminals connect to a device called a loop-wiring concentrator. This device forms a star-wired daisy chain. The concentrator has eight ports, seven of which connect to terminals. The seventh port connects to one of the host computer's communication channels. The system is twinaxial from the terminal to the wall plate, STP for the horizontal cable to the telecommunications closet, and a mixed STP/twinaxial from the concentrator to the host computer.

UTP: An active star hub allows the use of UTP. A balun (described below) allows UTP from the terminal to the hub and from the hub to the host computer. Each terminal must be treated as the last device in a daisy chain. One port must be terminated in its characteristic impedance. Allowable distances are:

Terminal to hub:	900 feet
Hub to host computer:	1000 feet
Terminal to host computer:	1000 feet

All three distance conditions must be met, so the total allowable distance is 1000 feet, not 1900 feet. If the hub is located near the host computer, the terminal can be located up to 900 feet from the hub. If the hub is over 100 feet from the host computer, the distance from hub to workstation must be shortened accordingly. If, for example, the hub is 600 feet from the host, the terminal can be only 400 feet away from the hub.

The IBM systems were originally designed to use coaxial or twinaxial cable:

3270 systems:	93-ohm coaxial cable
System 3x and AS/400:	100-ohm twinaxial cable

A twinaxial cable is similar to a coaxial cable, except that it has two center conductors in the dielectric.

To convert these coaxial/twinaxial-based systems to use UTP or STP requires a balun.

Baluns and Media Converters

Balun is a shortening of "balanced/unbalanced." UTP and STP require balanced signals, as described in Chapter 2. Coaxial cables carry unbalanced signals. A balun performs two functions: it converts signals between balanced and unbalanced modes and it provides any required impedance matching between the two circuits. In short, in going from coax to UTP, it converts a 93-ohm unbalanced signal into a 100-ohm balanced signal.

A balun consists of a transformer built around a doughnut-shaped toroid. Wires from the two interfaces are wrapped around the toroid. The number of windings on each side is proportional to the impedance to be achieved. For example, one side is a 100-ohm modular jack and the other side is a 93-coaxial connection.

Figure 5-23. Coax-to-UTP baluns *(Courtesy of AMP Incorporated)*

Baluns can be built into connectors and modular outlet inserts. The advantage of this over discrete baluns is the connectors and inserts maintain the proper polarity of the signals throughout the cabling system. Figure 5-23 shows baluns typically used for coax-to-UTP conversion. Figure 5-24 shows a balun that allows video to be converted for transmission on UTP.

Two related devices are the impedance-matching adapter and a media filter. An impedance-matching adapter allows cables of different impedances to be connected. For example, Token Ring hubs are available with either 100-ohm or 150-ohm ports. If you wish to connect a 100-ohm cable to a 150-ohm port, an impedance-matching adapter is required to provide a conversion between the two impedances without undue signal distortion and reflections. A media filter removes high-frequency components that may be part of the signal.

Figure 5-24. Video balun *(Courtesy of AT&T)*

PBXS AND TELEPHONE CABLING

Telephone wiring is also a form of network, although operating at slower speeds that present fewer problems in cabling. The slow speeds allow long runs on lower-grade cable. Even so, TIA/EIA-568A minimizes the differences between voice and data lines, distinguishing them by data rates (or frequency) rather than by applications. Most new installation and renovations use Category 5 for all horizontal cabling.

The heart of most telephone systems is the PBX or private branch exchange. The PBX is also know as a PABX (private automatic branch exchange) or CBX (computerized branch exchange). The PBX is an on-premises telephone switching system. It forms an interface between the telephone central office and the telephones in a building. In simplest terms, incoming calls go first to the PBX, which then routes them to the proper telephone. Modern PBXs, of course, are much more sophisticated, offering such features as voice mail, call forwarding, conference calls, and call accounting.

A PBX consists of a cabinet containing one or more shelves. Each shelf accepts plug-in boards that connect to a backplane in the back of the PBX. Different types of boards are available for different needs and levels of capability. Common types of boards in a PBX are:

- **Both-Way Trunk.** These are outside lines that allow both incoming and outgoing calls. Each trunk line connects to a port on the board. Thus an 8-port board can accept eight trunks.

- **Direct Inward Dial Trunk.** These boards accept incoming calls only and allow dialing of individual extensions from within the PBX.

- **Universal Trunk Circuit.** These boards mix both-way and direct-inward-dial capabilities.

- **Tie Line.** Tie-lines connect two PBXs without the need to place an outside call. Businesses use tie lines to connect different buildings without having to go through an outside line.

- **Digital Telephone.** These boards connect to digital telephones directly. Each board typically has 8 or 16 ports.

- **Analog Telephone.** Analog boards handle communications with analog phones. Like the digital telephone board, analog boards typically have 8 or 16 ports.

Telephone companies are increasingly using digital signals for carrying calls. With newer digital telephones and PBXs, it is possible to achieve end-to-end digital communications. The desktop telephone encodes your voice into a digital signal before transmission to the PBX. The PBX retains the digital form for transmission to the central office. Digitation can also occur in the PBX or at the telephone company's central office, so that portions of the transmission are analog.

It's important to know what portions of the building cabling are carrying digital signals and which are carrying analog. Requiring only a 4-kHz capability, analog signals can be carried on the simplest Category 1 cable. Digital signals operate at 64 kbps and require at least Category 2 cable. Category 3 cable is the preferred solution since Categories 1 and 2 are not recognized by premises building cabling standards.

CONVERGENCE

Convergence is one of the prevalent buzzwords floating around today. It refers to the blurring of telephony and computers. More and more, computers are taking on many of the basic capabilities of telephone systems, while telephone systems are adapting many of the characteristics of computers. PCs can be integrated with telephones to make many telephone functions easier to manage. Indeed, the computer may become the telephone. Computers today can offer voice mail, call management, and videoconferencing.

Both telephone and computer companies are working on methods to integrate computers and telephones. What's the point of all this? Consider these examples:

When a call comes in, your computer recognizes the number and automatically pops up a screen with notes from your last conversation with the caller.

You can highlight a list of names from a computer database and automatically fax a file to each. Or you can send the message by e-mail and even attach a voice message.

The cabling requirements of such computer telephony will vary. Both telephone and network cables will find use. While low-speed requirements can be met over telephone cabling and the PBX, high-speed needs will be met by the local area network. For high-speed needs, the PBX can become a node on the network. In any case, the movement is toward using Category 5 cabling for both telephony and data communications.

NETWORKS AND BUILDING CABLING

As we discussed earlier, the purpose of building cabling is to provide an application-independent infrastructure. With new networks, this is easy to do because recent networks use a star structure and UTP or fiber-optic cabling – fully compatible with TIA/EIA-568A and ISO/IEC 11801.

Accommodating legacy LANs like 10BASE-2 thinnet requires an *application-dependent* approach to cabling. You must be aware of the specific requirements of the application: what type of cable, connectors, topology, and so forth.

6   The Wired Building

We've looked at the main components of a premises cabling installation. This chapter looks at some additional issues involved in a building cabling system.

LINK STANDARDS

As important as it is to pick the right components to ensure the proper level of performance, components alone do not guarantee the required performance. They must be installed properly and work together. Does the assembled link meet performance requirements?

It's important to understand the differences between TIA and ISO standards for channels and links. A link is a specified portion of the horizontal cabling. A channel is the end-to-end run from one piece of equipment to another, say a network hub telecommunications closet to a personal computer in a work area. It is the path from transmitter to receiver and includes all cabling in the path. It can include both horizontal and backbone cabling. Both TIA and ISO models agree on the definition of a channel. The standards differ on what constitutes a link. The ISO model includes the patch cable and cross connect in the telecommunications closet, while the TIA model does not include them. TIA/EIA TSB-67, a testing standard discussed in Chapter 8, defines the basic link. Thus the TIA link is 90 meters maximum and the channel is 100 meters maximum – the same as the requirements for horizontal cabling. The ISO link is up to 95 meters long. It includes a patch cable, which can be up to five meters long, in addition to the 90-meter horizontal run. Figure 6-1 shows the differences between the two standards. These differences can affect testing since the link specifications are physically and electrically different.

Cases can be made for testing on either a link level or a channel level. Installers of a building cabling system are usually concerned with the link. The link forms the basic infrastructure for the building cabling. Since the cabling is performance driven, the installer is not concerned with what hubs or what workstations will be used. To a large degree, the cabling in the office or work area is out of the installer's hands. Cabling in the wiring closet or behind the walls is more secure and removed from users. This is not so for the work area cables. The installer, therefore, is concerned with the link, since this is the cabling most easily controlled during and after installation.

Realistically, however, the channel is the real measure of performance. The weakest link in the system is the two patch cables that fall outside the link model. Real-world performance runs from hub to workstation.

EIA/TIA-568A Model

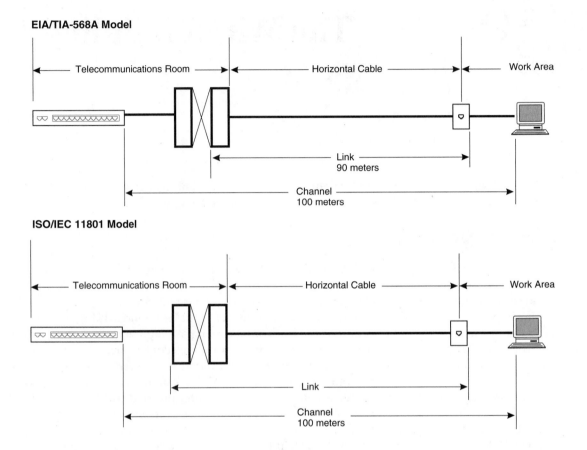

ISO/IEC 11801 Model

Figure 6-1. TIA and ISO link models

TIA link performance standards are based on Categories 3, 4, and 5 UTP and fiber. Coaxial and STP installations should be tested specifically to the requirements of the applications for which they are installed. As we mentioned, link performance is part of an informative annex rather than a formal part of the standard.

ISO/IEC 11801 defines five classes of applications:

- **Class A:** Voice and low-frequency applications. Cables are rated to 100 kHz.

- **Class B:** Medium-bit-rate applications, with cables rated to 1 MHz.

- **Class C:** High-bit-rate applications, with cables rated to 16 MHz.

- **Class D:** Very high-bit-rate applications, with cables rated to 100 MHz.

- **Optical:** Applications whose bandwidth is not a limiting factor in premises cabling.

Notice that Class C and D correspond to TIA/EIA Categories 3 and 5. ISO/IEC 11801 has nothing directly comparable to Category 4 links. In 568A, categories of cable and the links using them go by the same name. This similarity is a source of confusion. When your cable plant is tested to Category 5, does this mean the cable or the link?

In 11801, cable categories and link performance standards are clearly separated. Figure 6-2 shows the distances allowed for the different cables and classes. This is similar to Figure 2-25 for backbone cables for TIA/EIA-568A. The cable distances also apply to the backbone cable, so that you can calculate how far backbone cables can be safely run. Figure 6-3 shows the general relationship between TIA link categories and ISO classes.

Cable	Cable Distance by Class of Application (meters)				
	A	**B**	**C**	**D**	**Optical**
Category 3	2000	500	100	–	–
Category 4	3000	600	150	–	–
Category 5	3000	700	160	100	–
150-Ohm STP	3000	1000	250	150	–
Multimode Fiber	–	–	–	–	2000
Single-Mode Fiber	–	–	–	–	3000

Figure 6-2. Cable distances for ISO/IEC 11801 classes

Cable Type	Performance	TIA/EIA-568A Link	ISO/IEC 11801 Link
Category 1	100 kHz	–	Class A
Category 2	1 MHz	–	Class B
Category 3	16 MHz	Category 3	Class C
Category 4	20 MHz	Category 4	–
Category 5	100 MHz	Category 5	Class D

Figure 6-3. General comparison between EIA and ISO link performance levels

In Chapter 1 we talked about the levels of cross connects required in a building. Let's look closer at these and how everything fits into the system. While a cabling system deals primarily with the cable and related components like cross connects, patch panels, and connectors, a functional system must also consider other networks hubs and telephone PBX. While a performance-driven cabling system exists independently of specific network and telephony equipment, neither one can work without the other.

Like today's networks, the building cabling system uses a star configuration, consolidating cabling in different telecommunication closets.

A main cross connect is required as the center of every installation. The main cross connect can serve a single building or a campus of buildings. A horizontal cross connect is recommended for each floor of the building to act as a dividing point between the backbone and horizontal cabling. The intermediate cross connect exists to simplify larger installation by providing a cross connect between the main cross connect and the horizontal cross connect. In a campus application, for example, an intermediate cross connect can be used in each building. The main cross connect in one building feeds the intermediate cross connect in other buildings. (The main cross connect can also serve to distribute to a horizontal cross connect in the same building. Or it can feed an intermediate cross connect in the same building, or even in the same room.)

The main cross connect can also serve as a demarcation point between outside telephone cables coming into the building. This provides a clear separation between the telephone company's cable and responsibility, and the building's cabling.

While it's popular to show the main cross connect in the basement of a building, it can be placed anywhere in the building. In a high-rise building, placing the main cross connect in a middle floor can simplify cable routing. Similarly, a telecommunications closet in the middle of a floor, rather than on the outside, shortens the maximum distance any cable must be run.

A Note on TIA/EIA and ISO/IEC Terminology

While the TIA/EIA and ISO/IEC approach to building cabling is essentially the same, the terms they use to describe the parts of the system differ. Figure 6-4 provides a quick cross reference to these differences.

TIA/EIA	ISO/IEC
Main cross connect	Campus distributor
Intermediate cross connect	Building distributor
Telecommunications closet	Floor distributor
Backbone cable	Campus backbone cable Building backbone cable
Horizontal cable	Horizontal cable
Telecommunications outlet	Telecommunications outlet

Figure 6-4. Differences in TIA/EIA and ISO/IEC terminology

CROSS CONNECTS AND DISTRIBUTION FRAMES

The term *cross connect* usually refers to a passive system, one that does not include such devices as network hubs or routers. The cross connect is concerned solely with the wiring. When devices are added to the system, the term distribution frame is often used in place of cross connect. Distribution frame includes both the cabling interconnections – cross connects and patch panels – and the hubs, routers, and other devices. Distribution frames have the same hierarchy as cross connects: main distribution frame, intermediate distribution frame, and horizontal cross connect.

HOMERUN OR COLLAPSED CONFIGURATION

The collapsed or homerun configuration consolidates all equipment in the main cross connect. With fiber-optic cabling especially, this configuration can simplify the cabling plant. Because fibers can easily be run hundreds of meters, they can easily run from the main cross connect directly to the work area. In a small building, copper cables can also be used in a home-run configuration, but large buildings will require telecommunications rooms on one or more floors.

One reason for the attractiveness of the telecommunications closet is that network equipment like hubs will be located there. The 100-meter limit on copper cables requires that hubs be positioned locally, either on the same floor or an adjacent floor. With fiber, the hubs can be located a great distance away. The advantage is to center all equipment at a single location to make any network changes easier to accomplish. This centralized administration simplifies moves, adds, and changes.

Even if an all-fiber installation is chosen for data needs, telephone wiring must also be considered. The low-speed requirements of telephony allow longer runs when necessary and place fewer restrictions on the cable used. The ISO requirement for a Class A link for voice circuits allows runs of up to 2000 meters for a signal up to 100 kHz.

Fiber proponents argue that the homerun configuration simplifies the cabling and increases reliability by reducing the number of interconnections in the run to the desktop. The main drawback is that it can violate TIA/EIA-568A and ISO/IEC 11801 requirements for a horizontal cross connect on each floor to separate horizontal and backbone cabling. In some cases, the homerun configuration can use the same cable for both the horizontal and backbone run. Still, a cross connect on each floor should be added to satisfy requirements. The TIA/EIA is currently studying a proposal for fiber that will allow homerun configurations with a combined backbone/horizontal cabling distance of 300 meters (Figure 6-5). The requirement for a horizontal cross connect will be eliminated. Either the cable can be run directly from hub/patch panel in the equipment room to the work area outlet or a splice can be used to separate the backbone cable from the horizontal cable. The horizontal cross connect becomes unnecessary since all rearrangements are made in the equipment room.

Collapsed Networks

A variation of the homerun system used in many networks is called the *collapsed backbone*. In this case, all the equipment is not necessarily in the main distribution frame. All hubs are connected through a central hub or router in the main distribution frame. The hub or router connects in a star configuration to hubs in the telecommunications closet on each floor. The advantage of this is to make the network on each floor a subnetwork of the larger network. A router keeps all local data local within the subnetwork. If a message is intended for a station on another floor, the router passes the message only to the station on the destination floor.

Such a collapsed backbone increases efficiency by reducing the number of stations that a message goes to in a shared-media network. Consider a building, for example, where the marketing, sales, engineering, and accounting departments of a company each are on a different floor. For the engineering department, most network communications are within that department. A lower percentage of communications occur with other departments. In a large shared-media network like Token Ring or Ethernet, the messages still pass to every station on the network. If the network is configured as a single large network, the traffic on the network increases and the response time decreases. By using a router-based collapsed backbone to form subnetworks, network efficiency increases. Messages between stations in the engineering department are confined to the engineering department; only those destined for other departments pass through the router to the destination department.

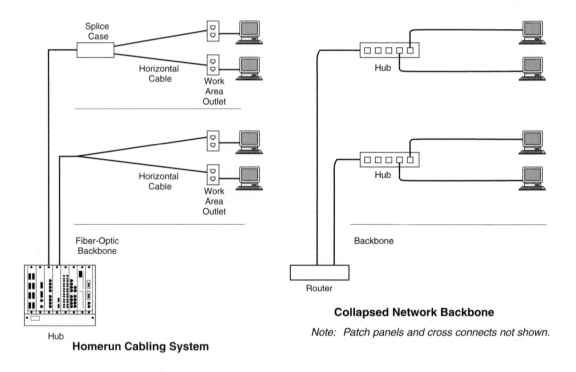

Figure 6-5. Homerun cabling system and collapsed network backbone

FAULT TOLERANCE

Fault tolerance refers to the ability of a system to continue to operate if an error occurs in part of the system. The error should not crash the system. Fault tolerance does not mean the system's operation is not diminished; it usually means the problem is isolated and the rest of the system operates. Total fault tolerance is possible so that no system functionality is lost, but this often is too costly to be practical.

In planning for fault tolerance, you must judge how critical each part of the premises cabling system is to your business. Any disruption is irksome and expensive, but some disruptions are much worse than others.

The main approach to fault tolerance is redundancy. Subsystems are duplicated in whole or in part. A backbone cable between buildings can be redundant. But how redundant is redundant? Take a fiber-optic backbone requiring four fibers. You have several choices:

- Run eight fibers in a single cable. If any of the four active ones fails, you can use one of the spares.

- Run two separate 4-fiber cables in the same conduit. You have a spare cable if the first breaks.

- Run two separate 8-fiber cables in the same conduit. This combines the first two approaches.

- Run separate cables in *separate* conduits. Then if an entire conduit fails, you still have a backup.

- Use a mesh structure. Even separate conduits run between the same two buildings can fail. If one building burns down, both links are rendered useless. A mesh structure interconnects each building to each building. Now the main conduit runs from B to A, the backup from C to A and so forth.

- Combine redundant building-to-building conduits with a mesh structure.

You can see that the fault-tolerant structure becomes increasingly complex – and expensive to implement – as you increase the degree of fault tolerance from spare fibers in a cable to a completely redundant mesh system. Figure 6-6 shows examples of backbone fault tolerance.

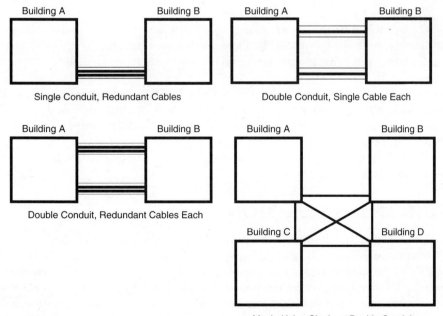

Figure 6-6. Backbone fault tolerance

Going beyond the building-to-building backbone, fault tolerance in the backbone cable, horizontal cable, and in the interconnections all can be accomplished to varying degrees. Again, this is done through redundancy. As shown in Figure 6-7, fault tolerance is often achieved by having duplicate telecommunications closets feeding the horizontal cable to the work area.

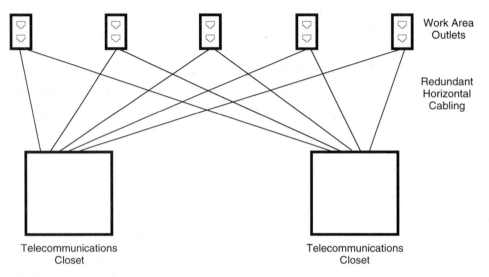

Figure 6-7. Horizontal fault tolerance

The backbone cable is typically seen as the critical link since its interruption will affect the highest number of users and is probably the most time-consuming and expensive to repair. On the other end, a work area cable is a small blip on the screen of critical concerns. It's easily and quickly replaced.

Many network devices have fault tolerance built in. FDDI, for example, can wrap the ring from the primary to the secondary to avoid a cable break. Some hubs allow redundant back-bone links, with automatic switchover from the active to the standby cable if a failure occurs. Other networks offer methods of rerouting traffic to avoid points of failure. The long-distance telephone service does the same thing by providing multiple routing paths. If one path goes down, many others are available.

PLANNING AN INSTALLATION

A fundamental issue in planning a building cabling structure is to first decide what to bring to each work area. How many outlets and of what type are needed? A single data connection? Multiple data connections? How many phone connections? Are phone connections for analog or digital telephony? What level of performance is required today? Tomorrow?

While overspecifying the number of connections in a work area and the number of cables running to it will increase the cost of an installation, it is still cheaper than pulling new cables at a later date. Modular outlets allow the flexibility of changing the outlet interface but don't solve the problem of pulling new cable. Therefore, it is smart to anticipate future needs and pull the cable to the outlet at the very least.

The premises cabling has the longest life cycle of any part of the network or telephone system. Software has the shortest life cycle, with frequent upgrades and changes. Personal computers can become obsolete in increasingly short cycles as the demands of evermore sophisticated software places greater demands on the horsepower of the PC. Network equipment, too, may require upgrading and changes to accommodate newer demands and sophistication. But the premises cabling system should have a minimum useful life of 10 to 15 years. Therefore, take great care in planning the system. Forethought is much more cost-effective than recabling.

- Figure on one work area for every 100 square feet of floor space.

- For each work area, use a minimum of two connections: one for data and one for voice. Two voice and two data connections ensure greater flexibility. Also consider whether you should provide for a direct central-office connection for any voice lines, bypassing the PBX.

- Cable everything. Consider, for example, data/voice outlets for every seat in a conference room and in reception areas. The brave new world of multimedia, video conferencing, portable computers, and anytime/anywhere communications means that you should anticipate the future need to plug into the information infrastructure.

- The telecommunications closet should be located at a common vertical access. Ideally, it should also be located where all work areas are within 100 meters. A single closet can serve an area of about 10,000 square feet. Multiple closets may be required if the area exceeds this. Research by AT&T shows that 99% of the time, workstations are within 93 meters of the closet.

- Carefully evaluate the advantages and disadvantages of cross connects versus patch panels. Cross connects are less expensive, but harder to maintain except by experienced persons. Do you anticipate many changes? Do you have a qualified staff? Patch panels can be easily used by network administrators, who have the in-depth knowledge of the cabling topology but little skill at terminating cable in punch-down blocks. A compromise often made is to use cross connects for voice and patch panels for network data.

- Be generous in anticipating future needs.

- Approximate needs for a telecommunications closet are based on the area served and based on the recommendations of TIA/EIA-569:
     10,000 square feet: 10 X11-foot closet
     5,000 square feet: 10 X7-foot closet

- Equipment can be contained in the telecommunications closet or in a separate equipment closet. The choice depends not only on preference, but on security concerns and the needs of the equipment, including the power and air-conditioning requirements. The equipment room includes the PBX, network hubs, network servers, minicomputers, and so forth.

- Don't try to get double duty out of a cable by combining different signals. While it is possible to use different pairs for voice and low-speed data on a cable (10BASE-T and voice, for example, require only three pairs total), terminating them in separate outlets, the practice is to be avoided.

- Leave extra cable slack in the walls. While there is a minimum amount recommended for rewiring outlets, a greater amount will provide more flexibility. For example, leaving a 6-meter loop in the ceiling or floor will not only permit an outlet to be rewired, it will allow the outlet to be moved. Remember that offices are rearranged and furniture moved. Don't allow the positions of the outlets to decide the arrangement of offices. In particular, poorly placed floor outlets can be a hazard to flexible office arrangements.

- The premises cabling system should be laid out using scale drawings. Cable distances should be calculated on the actual routing of cables, not on straight-line distances.

- Consider using factory-terminated cable assemblies in the system. This is especially true for patch cables, but can also apply to the "behind-the-wall" cables in horizontal and backbone runs. While the component costs may be higher, installation costs may be lower. What's more, you receive assemblies that are tested and ready to install. With careful installation (avoiding the pitfalls discussed in the next chapter), the cable plant should be quickly and easily brought up and running. Cable assemblies require much careful planning if custom lengths are to be used. You can't make on-the-spot adjustments if the assembly is six inches too short.

- Consider voice and data together. There is a tendency for telephone people and network people to view their domains exclusively and separately. This view is increasingly short-sighted. First, voice and data communications are merging fast. Second, PBX makers are (somewhat belatedly) offering ways to carry high-speed network data and even serve as part of the network. Third, ATM is both a telephony and network application. More and more, you should be careful to make sure the network and telephone people are talking and understanding the needs of the other.

CHAPTER **7**

Installing
Cabling Systems

The most common mistakes in installing a premises cabling system involve either improper cable handling or improper terminations.

In general, proper cabling handling becomes increasingly important with better grades of cable. For instance, Category 5 UTP and optical fibers are more sensitive to improper installation than Category 2 or 3 cable. To avoid problems, installation practices should be based on the requirements of Category 5 UTP and fiber. Doing so avoids the need to have multiple procedures based on the grade of cable. Even with this recommendation, there are still differences that must be considered in dealing with copper and fiber.

Cables can be run in raceways and conduits installed in open spaces. A raceway is an enclosed path through which the cable is run. A conduit is a circular raceway. They can be run horizontally in the floor and ceiling as well as vertically. If a cable is run through an air-handling space, it must either be enclosed in a raceway or be plenum rated.

A pathway is the path between two spaces in a premises cabling structure. A space can be a wiring closet or a work area; the pathway connects them. The pathway itself can be structured – a raceway running under the floor or through a modular office wall. In some cases, cables are simply laid on the supports of a suspended ceiling in an unstructured manner, although this practice runs counter to the requirements of TIA/EIA-569.

A number of conduits and raceways are available for cable. If the run is long enough, the raceway will have intermediate access points used to pull the cable during installation. Such pull boxes are necessary to keep tensile loading on the cable within specified limits.

GENERAL CABLE INSTALLATION GUIDELINES

Neatness Counts

The premises cabling plant should never resemble a rat's nest of disorganized wiring. Cables should be organized and dressed with cable ties to keep them neatly bundled. If the need arises to troubleshoot or revise the cabling, a neat system is much easier to work with. Even so, neatness shouldn't be carried to the extreme where it stresses the cable in unacceptable ways (as described below).

Never Exceed the Minimum Bend Radius

All cables have a minimum bend radius. If this radius is exceeded, performance degrades. Even straightening the bend may not return the cable to its original performance level.

Cable should be installed in swept curves, avoiding sharp bends. Be careful when pulling cable that the conduit does not cause excessive bends or kinks. If necessary with conduits, use a fitting that increases the radius of the turn (Figure 7-1).

NO! Tight Radius   YES! Gentle Radius

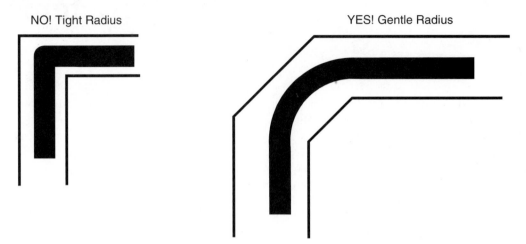

Figure 7-1. Use proper fittings to maintain swept bend radius

Too much attention to a neat appearance in a wiring closet can also cause the bend radius to be exceeded. Don't bend cables sharply simply to keep them close to equipment.

In the work area, cable is often stuffed into the outlet box. Be careful here to maintain the radius and avoid sharp bends and kinks. Some outlets have built-in features to allow a generous radius to be maintained.

Hardware such as cross connects, patch panels, and fiber-optic distribution boxes often offer cable management features. Figure 7-2 shows a fiber-optic distribution box. The box uses a tray to organize the fibers neatly and orderly, maintaining a generous bend radius on each fiber. Such boxes typically allow fibers to be connected either by splices or connectors.

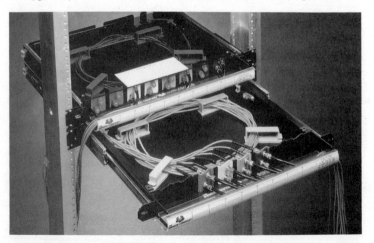

Figure 7-2. Fiber-optic distribution box demonstrates the cable management features of hardware *(Courtesy of The Seimon Company)*

Optical fiber cable has two minimum bend radii: one to be observed during the installation procedure and one for long-term use. During installation, the tensile loads of pulling necessitate a greater radius. The installed cable can be wrapped in a tighter radius.

Here are general rules for the minimum bend radius:

UTP: 4 times the cable diameter for 4-pair horizontal cable; TIA/EIA-568A recommends a 1-inch minimum radius.

10 times the cable diameter for multipair cable

Fiber: 4 times the cable diameter (about 1.8 inch for a zipcord construction)

Maintain Proper Tensile Loads During and After Installation

Tensile load refers to the stresses exerted on the cable during pulling or after installation. A cable's tensile strength rating refers not to the failure point of the cable – such as pulling so hard the conductors break – but to a reasonable limit above which performance degrades or the long-term reliability is impaired.

Too much tensile loading on a twisted-pair cable, for example, can stretch the twists, changing the geometry of the cable. With Category 5 cable, whose performance depends on maintaining the proper geometry, exceeding the tensile load can ruin the cable for 100-MHz performance. Too much tensile loading is a quick way to turn a Category 5 cable into a Category 3 cable – permanently.

The tensile load on an optical fiber involves its long-term reliability. Overstressing a fiber can cause tiny flaws to grow, possibly leading to ultimate failure of the fiber. Fortunately, the cable's construction, especially the strength members, significantly decouples the fiber from pulling forces. Despite the reputation of fiber for being fragile, fiber-optic cables can withstand higher tensile loads than copper cable.

Tensile loads should be monitored during installation. Problems occur during pulling from several sources:

- Pulling too long a cable
- Pulling around corners
- Pulling over obstacles
- Pulling through a conduit too small for the pull being performed
- Not observing the recommended loading limit for the type of cable being pulled

Cables should be well supported, especially in hanging applications, to ensure that after-installation tensile loadings are not exceeded.

Common UTP cable (24 AWG, four-pair cable) has a tensile loading limit of 25 lb. For other conductor sizes, use the information in Figure 7-3, which shows the maximum tensile loading for a given conductor size. Multiply the maximum tensile strength by the number of conductors in the cable. Notice that solid and stranded conductors of the same wire gauge have different load limits.

A typical two-fiber optical cable for horizontal application has a tensile load of 100 lb during installation and 14 lb after installation; other two-fiber cables can offer pulling loads ranging up to 210 lb for installation and 65 lb long term.

Conductor Size (AWG)	Type	Maximum Loading (lb)
26	Solid	2.0
	Stranded	2.2
24	Solid	3.2
	Stranded	3.6
22	Solid	5.0
	Stranded	5.6

Figure 7-3. Loading forces for various conductor sizes

Always follow the manufacturer's recommendation on loads.

Cables can be hung from J or T hooks or bridle rings above suspended ceilings. Make sure the hooks are spaced closely enough to keep the tensile load of drooping cable within specified limits. TIA/EIA-569 specifies a maximum spacing of five feet, but shorter spacing may be required. If cables are run in trays or conduits, be careful that heavier cables don't load lighter cables by lying atop them.

Observe the Vertical Rise of Riser Cables

Vertical cables must be well supported. Since a cable rising vertically must support its own weight, it must be supported at regular intervals. Starting at the top of a building and pulling downward allows the weight of the cable to help the pull.

Avoid Deforming the Cable When Supporting It

Cables should not be deformed by fastening hardware. Cable ties, straps, clamps, and staples can compress the jacket and deform conductors. Again, such compression can degrade performance, often irreparably. Always use care to prevent overtight placement of support hardware. Be sure cable ties aren't overcinched and that Velcro straps and nail-on clamps are snug without being overtight. Staples, for both hand and gun application, are available that bridge the cable without crushing it. Always leave staples snug enough to support the cable but loose enough not to stress it.

For vertical applications, split wire mesh grips support cable without crushing it. The mesh spreads the holding force over a wider area to prevent overstressing a single point on the cable.

Keep Cables Away from EMI Sources

Because power cables can interfere with signals on copper cables, they should be kept separated. If it is necessary to run power and signal cables in the same conduit, separate inner ducts must be used for each.

Fluorescent lights, radio frequency sources, large motors, induction heaters, and arc welders also present serious sources of EMI that must be considered in routing cable. Closed metal conduits generally provide sufficient shielding of most sources. If the noise source is inductive (large changes in current), a special ferrous induction-suppressing conduit may be required. Fluorescent light fixtures represent the most common noise source in offices. Keep cables at least five inches away from any fluorescent fixtures.

Open or nonmetallic pathways offer no shielding from noise sources, therefore cables must be separated from them. Figure 7-4 shows TIA/EIA-569 recommendations for separation distances for power wiring above 480 V.

Condition		Minimum Separation Distance (inches)		
Power Line Pathway or Electrical Equipment	Signal Line Pathway	Less than 2 kVA	2 - 5 kVA	Over 5 kVA
Unshielded	Unshielded	5	12	24
Unshielded	Shielded	2.5	6	12
Shielded	Shielded	–	3	6

Shielded: Metallic, grounded conduit

Unshielded: Open or nonmetallic conduit

Figure 7-4. Recommended separation from high-voltage power sources

Separation distances are not a concern with all-dielectric fiber-optic cables. These are immune to electromagnetic interference. For hybrid optical cables that also contain copper conductors or metallic strength members, the recommended separation distances should be observed.

Observe Fill Ratios for Conduits

The cross section of a conduit should be only 50% filled with cable. This ratio applies to each inner duct of a conduit. Keeping a low cable-to-conduit area ratio helps reduce pulling loads.

Reduce Loads by Pulling from the Center

Most often cable is pulled from one end to the other, that is, from the telecommunications closet to the work area. If you think tensile loading will be exceeded because of any reason – long pulls, obstructions, bends in the path – you can start at a midpoint and pull half the cable in one direction and half the cable in the other direction. This effectively reduces the loading over that of a single pull.

Avoid Splices

Cable should, whenever possible, be run in a continuous length, without splices. This splice-free recommendation applies especially to horizontal cable. With some backbone applications, splicing may be necessary. Always locate splices in an accessible point, never behind walls or above inaccessible ceilings. Any mechanical terminations – connector or splice – is a potential point of problems that must be left accessible.

One possible exception to this splice-free recommendation is the 300-meter fiber homerun system. Splices are sometimes used in place of a disconnectable horizontal cross connect. All rearrangements of cables can be done in the equipment room where hubs are located. Since rearrangements are not made at the horizontal cross connect, splices can be used without degrading the flexibility of the system.

INTERBUILDING CABLES

Cables run between buildings can be directly buried, run through underground conduits, or hung aerially on poles. Underground conduits are the favored method in most cases and are used in around 80% of all premises cabling interbuilding applications. For long runs over a mile, where the cost of conduit becomes prohibitive, direct-buried cable is a reasonable alternative. Direct-buried cable requires armor that can withstand environmental extremes of temperature and moisture and the hazards of ravenous rodents. Since fiber is the preferred medium for interbuilding runs, we concentrate on it.

While fiber and copper share common pulling practices, fibers can also be blown through conduits. First, a *tube cable* is laid. This consists of an outer jacket surrounding up to 19 inner tubes. Each inner tube accepts a single fiber bundle of up to 18 fibers. The blowing device bleeds compressed air into the inner tube to float the fiber, which is paid off a motor-driven reel at a rate of about 100 feet per minute.

Air-blown fiber depends on continuous, splice-free runs, which typically is not a problem with premises cabling applications. One advantage of air-blown fiber is that it is easy to blow new fibers through empty tubes if it becomes necessary to expand capacity. Damaged fibers can also be easily blown out of the tube and new ones installed.

Most outside plant fiber uses a loose-tube design in which the fibers "float" within the tube. Gel fills the tube to prevent moisture from forming on the fibers. Burial can be either by trenching or plowing. In trenching, a trench is dug, the cable is placed in it, and the trench is filled. Plowing uses special cable-laying equipment to dig a ditch, lay the cable, and cover it in one operation. It's advisable to place the cable below the frost line in your area.

Conduit provides greater flexibility than direct burial. First, it makes repairs and installation of new cable easier since no additional digging is required. Using inner ducts to segregate cables makes pulling new cables easier by eliminating snags, frictions, and blockages that can occur when new cable is installed over old. Inner ducts can also be color coded by function. Some installers place pull ropes or tapes in empty ducts to make future installations even easier.

TERMINATING CABLES

Misapplied connectors can be a source of problems in premises cabling. Particularly with Category 5 UTP and fiber, careless application can degrade performance considerably. Again, a reminder of an earlier statement: most network problems occur at the physical layer, with the cables and connectors.

Terminating Modular Connectors

Figure 7-5 shows the basic procedure for applying a modular plug to a connector. The general termination procedure is simple:

- Strip away the cable jacket to expose the individual pairs.

- Untwist the pairs just enough to insert them into the plug – no more than ½ inch for category 5 cable and 1 inch for Category 4.
- Insert the cables into the plug, being careful to use the correct order for the wiring pattern being used.
- Cycle the crimping tool.
- Inspect the termination. The clear plastic of the plug allows you to inspect the termination area and check that the pairs are in the right order.

The importance of carefully inspecting each termination cannot be overemphasized. Careless or inadvertent terminations are a major source of problems and errors in premises cabling.

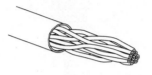

1. Strip the cable jacket

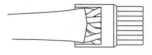

2. Assemble the load bar on the conductors

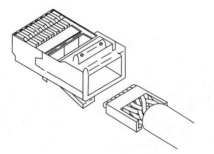

3. Insert cable in connector housing

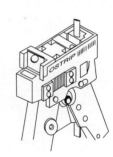

4. Crimp

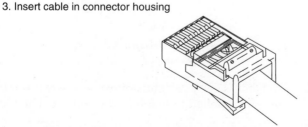

5. Inspect the finished termination

Figure 7-5. Terminating a modular plug

Figure 7-6 shows a typical termination procedure for terminating a modular jack with 110-style punch-down blocks. The wires are carefully positioned in the contacts (taking care to limit untwisting of pairs) and then pushed into place with a punch-down tool.

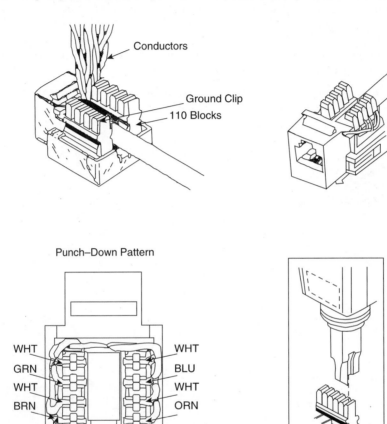

Figure 7-6. Terminating a 110-style modular jack *(Courtesy of AMP Incorporated)*

Terminating a cable in a punch-down block is fairly straightforward. The wire is positioned in the IDC contact and the punch-down tool is applied to seat it in the contact. The tool also cuts off the end of the wire.

Here are some things to pay attention to:

- Remove as little of the outer jacket as possible.

- Maintain the twist on the cables as close to the connector as possible. *With Category 5 UTP, do not untwist more than 0.5 inch.* For Category 4 cable, the limit is one inch.

- Make sure all components are rated to the category being installed.

- Use the same cabling scheme throughout the installation. That is, terminate each plug in the same wiring pattern. The preferred pattern is T568A or T568B. Don't, for example, mix T568A and T568B patterns. Using a mixed wiring pattern will only lead to future confusion and incompatibilities.

- Crossover cables and connections, such as those required to connect two hubs (see Figure 5-11) should not be part of the premises cabling system. These should be treated as special cases outside of the system. Trying to incorporate them into the cabling plant leads to long-term confusion and incompatibilities.

- Pay close attention to the position of each conductor within the connector. The pattern should be the same at both ends of the cable.

- While a connector can be reused, it's often faster and more cost-effective to cut off and discard an incorrectly applied connector.

- Make sure you have the proper connector for the cable. Some modular plugs are designed for stranded conductors, while others accept solid conductors. Some plugs accept both stranded or solid conductors.

- Maintain proper cable-management practices with regard to tensile loads and bend radii. Neatness counts, but not at the expense of degraded cable.

Factory-terminated patch cables are preferred in the work area and at cross connects. This is especially true for Category 5 cables, which are more prone to misapplication. Single-ended cables are available terminated at one end with a modular plug and unterminated at the other end. The unterminated end can be punched down at the cross connect.

Terminating Fiber-Optic Connectors

Earlier, in Chapter 4, we showed a typical installation procedure for a fiber-optic connector. Here are some useful things to remember about working with fiber:

- Use epoxyless connectors to speed application time. Eliminating epoxy cuts the number of consumables, cuts the potential for mess, and slashes installation time. The lack of epoxy also speeds polishing time since epoxy is harder to remove than glass. The connectors have proven themselves reliable and easy to use.

- Be careful with the fiber ends removed from the connectors. They can puncture the skin like a splinter and are hard to remove. Some installers use a piece of tape to collect them.

- Cleanliness is essential. Keep the fiber clean and dry when working with it. Dirt, films, and moisture are the enemies of fiber performance and reliability. Likewise, keep tools and supplies clean so they don't contaminate fibers.

- Limit the number of passes with the stripping blade. One pass is recommended. Then clean the fiber with a lint-free pad soaked in pure (99%) isopropyl alcohol. Isopropyl alcohol of less than 99% purity can leave a film on the fiber.

- Maintain the proper bend radius. Most outlets and organizers have built-in fiber management features that will allow generous radii to be achieved.

- Leave slack at each interconnection, such as inside wall and floor outlets and organizers. At least one meter of fiber is recommended. This allows for future repairs or rearrangement of cabling.

Polarization of Fiber-Optic Cables

TIA/EIA-568A recommends duplex SC connectors as the standard interface. The connectors are labeled A and B. This standard requires a crossover at every coupling adapter. At a wall outlet, for example, the in-wall A-connector mates with the work area B cable, as shown in Figure 7-7. The crossover simply means that one end of a specific fiber is labeled A and the other end is labeled B.

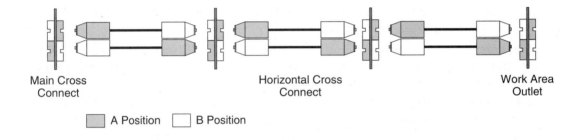

A Position   B Position

Figure 7-7. Polarization of duplex fiber-optic cables

The SC system is polarized by a key on the connector and keyway on the adapter. It becomes important to maintain polarization throughout the installation so that transceivers on equipment are always properly connected. Adapters at each end of a cable run should be installed in the opposite manner from each other. In other words, if the adapter keyway is up at one end, it should be down at the other end.

For multifiber cables, pair even-numbered fibers with odd-numbered fibers. Fibers 1 and 2 form a pair, 3 and 4 form a second pair, and so forth.

This may seem overly complex, but it has a purpose. Consider two electronic devices, for example a hub and a NIC. For devices, the receiver port is A and the transmitter port is B. If the NIC connects directly to the hub, the transmitter port (B) of the NIC connects to the receiver port (A) of the hub. In other words, port A at one end connects to port B at the other end. What the 568 rules do is to maintain this A-to-B and B-to-A orientation at every adapter throughout the system. By thoughtfully ensuring polarization throughout the system, you save users from having to think about it later.

GROUNDING

Grounding and bonding is important to the cabling system. Grounding involves establishing a low-impedance path to earth. Grounding has two purposes. Most important is safety to prevent the buildup of voltages that can result in hazards to persons or equipment. The second purpose of grounding is to help preserve the signal transmission quality.

Structural steel in a building can be used for ground, or a dedicated ground riser can be installed. The riser is dedicated to the telecommunications and network equipment and is separate from the grounds in electrical circuits. As such it is not intended to carry fault currents or perform safety functions associated with the building's power system. A second conductor, the coupled bonding conductor, can be installed in parallel to the telecommunications riser grounding bar to provide surge protection. While the telecommunications grounding system is separate from the electrical grounds, they are bonded together at the service entrance.

In the work area, equipment is grounded to the power grounding conductor. Any surge protection equipment is likewise grounded to the local power ground.

Earth ground is important with some electronic equipment to prevent ground loops. A ground loop occurs when two separate pieces of equipment are grounded without reference to earth. They are not connected to a common ground. Although grounded, their ground potentials may be different. Problems arise when shielded cables are run between the two pieces of equipment. The shields are grounded to equipment at each end, but because their potentials are different, a voltage difference exists between them allowing noise currents to flow in the shield. Do not rely on equipment chassis ground to provide an adequate, noise-free ground termination for shielded cables. The equipment should be grounded to the equipment rack, and the rack should be grounded to the telecommunications earth ground system. Unshielded interconnections do not require grounded equipment.

Bonding is the method used to ensure that the safety path to earth has electrical continuity and the capacity to conduct any currents that occur. Bonding is the method by which electrical contact is made between metallic parts of the system. Bonding is important because it determines the quality of the ground system. Poor bonding results in joints that exhibit high impedance and discontinuities to the flow of current. It also allows the buildup of differences in potential in the ground system.

Bonded joints should be joined by welding, brazing, swaging, soldering, screws, and so forth. Attention must be paid to the surfaces of the metal to be joined. The surfaces should be bare metal, cleaned of any nonconductive coating such as paint or grease. The joint should also be protected against corrosion, which decreases conductivity. It's a good idea to seal the bond.

There are three standards widely used for grounding and bonding:

- National Electrical Code has grounding requirements that are the basis of local building codes. The NEC is concerned with safety issues and requires the earth ground.

- TIA/EIA-607 – *Commercial Building Grounding and Bonding Requirements for Telecommunications* – presents a general overview of a grounding system for buildings.

- IEEE P1100, known as the Emerald Book, covers powering and grounding for sensitive electronic equipment.

Metallic cable sheaths must be grounded at the building entrance. The Emerald Book recommends grounding telecommunications cables to structural steel; TIA/EIA-607 requires bonding them to a grounding bar in the entrance space. The grounding bar, in turn, is bonded to the power service equipment. The advantage of 607 is that it is compatible with buildings that don't have structural steel. It also important to note that telecommunications cables should not be bonded to any structural steel members that are used for lightning protection.

In short, the grounding system must provide a low-impedance (less than 25 ohms) path to earth. In addition, the grounding conductors must be well bonded and sufficiently sized to conduct any anticipated currents.

SURGE PROTECTION

Surge protection isolates the cabling system from damage from surges in voltage and current from lightning, power faults, and electrostatic discharges. Buildings have primary protection built in by codes to protect people and property. But the primary protection may not be fast enough to protect sensitive electronic circuits. Nor do they often provide sufficient clamping to prevent damaging voltages or current from reaching the electronics. Clamping refers to the how quickly and to what level the protective device shunts surges to ground. With electronic circuits, surge energy must be dissipated quickly so that the total power passing through the protection device is limited.

Secondary protection can be designed into the premises cabling system. Surge protectors can be included at cross connects and outlets. It's a good idea to install protection as close as possible to the equipment being protected.

CHAPTER 8

Certifying the Cabling System

After the cabling system is installed, it should be tested and certified so that it meets the performance specifications. Testing and certification have different implications in your building cabling. Testing implies that certain values are measured. Certification, on the other hand, compares these measured values to standards-derived values to see if the values are within the limits specified. If you want a Category 5 UTP cabling plant according to TIA/EIA specifications, the certification process ensures that all relevant characteristics meet the TIA/EIA Category 5 requirements. Thus testing is a quantitative procedure, while certification is both quantitative and qualitative.

Certification is especially important for high-speed applications. Simply installing Category 5 UTP cabling does not guarantee 100-MHz performance. The pitfalls of an incorrect installation are many, but testing will easily identify most. An understanding of the tests and their significance will speed troubleshooting if problems arise.

COMPONENT, LINK, AND CHANNEL TESTING

We've emphasized throughout this book the need to use components with the proper rating throughout the cabling system. Failure of even part of the system to meet Category 5 performance, for example, could result in the entire link failing to allow 100-MHz performance. Likewise, having all components rated to Category 5 does not guarantee performance. Installation errors – too much untwisting of pairs in a connector, the wrong modular jack or plug for the wire being used, overstressed cables, cable ties too tight, and so forth – can degrade performance. It is important, therefore, to test the assembled system.

TIA/EIA TSB-67

Telecommunications System Bulletin 67 covers field testing methods for four-pair twisted-pair cabling systems on both the link and channel basis. The standard was written to overcome confusion and shortcomings in TIA/EIA-568A. The standard defines performance requirements, required tests, and accuracy limits in test equipment.

TIA/EIA-568A defines component requirements; TSB-67 specifies performance requirements for installed links and channels. Most test equipment performs a generous suite of tests to give you a full picture of the parameters of the cable plant. TSB-67, however, only requires four tests: wire map, length, attenuation, and NEXT loss. TSB-67 does not replace 568A; it simply provides testing requirements that are compatible with 568A.

CERTIFICATION TOOLS FOR COPPER CABLE

Of the many instruments available for testing twisted-pair and coaxial cable, the cable tester is the most versatile for testing and certifying a cable plant. A cable tester is a handheld unit that can perform the variety of tests required for certifying a cabling installation.

Figure 8-1 shows a typical cable tester. Since some tests require measurement of a signal, a signal injector is also required. Signal injectors are used at the opposite end of the cable from the cable tester. Since measurements like NEXT must be performed at both ends of the cable, it is necessary to switch the tester and injector. Some newer injectors allow measurements for both ends of the cable to be performed without swapping the tester and injector. The tester automatically controls operation of the injector.

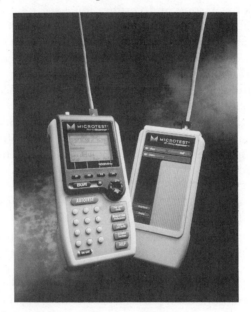

Figure 8-1. Cable tester and signal injector *(Courtesy of Microtest)*

A typical tester can perform the following tests:
- Wire map (required by TSB-67)
- Length (required by TSB-67)
- Near-end crosstalk (required by TSB-67)
- Attenuation (required by TSB-67)
- Attenuation-to-crosstalk ratio
- Impedance
- Capacitance
- Loop resistance
- Noise

The sophisticated capabilities of testers give you great flexibility in the tests performed. The tester will, for example, allow you to perform all tests automatically – autotesting – or to perform specific tests. And to make testing even easier, high-end testers have a built-in database of values to be tested. Even further, you can specify the performance criteria to which you want to test. The tester shown in Figure 8-1, for example, will automatically test to 10BASE-T, 100BASE-TX, 100VG-AnyLAN, 4- and 16 Mbps Token Ring, TP-PMD, 51 and 155-Mbps ATM, TIA/EIA-568A Categories 3, 4, and 5 links, ISO Classes C and D, and TSB-67 channels and links. You can test to either application- or performance-driven criteria. The tester maintains the database of specifications for different applications in a type of memory called flash memory. You update the database by downloading data from a PC into the tester's flash memory.

You can also define custom tests so that special cases or new standards can be included.

Figure 8-2 shows general setups for link and channel testing. The basic difference between the two tests is whether patch cables to equipment are included.

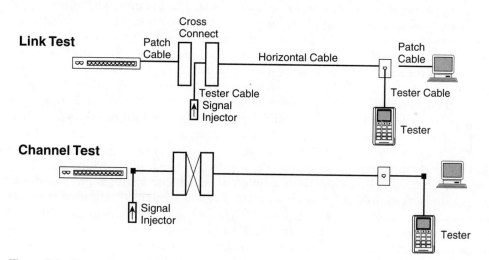

Figure 8-2. General setup for link and channel testing

Wire Maps

A fundamental test is the continuity test. Is there a circuit from end to end on each wire? Absence of continuity often indicates a broken wire or faulty connector termination. In a wire map test, the tester checks whether the pairs are properly terminated in the proper connector positions at each end of the cable. A common error is to cross pairs as shown in Figure 8-3. The tester will display the pairing of the cable, including the ability to detect shorts, opens, and crossed pairs. It can test against specific wiring patterns, such as T568A or T568B, and display errors.

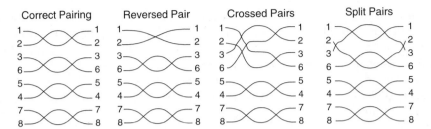

Figure 8-3. Continuity test for crossed, reversed, and split pairs

Cable Length

A cable tester can measure the length of the cable by time-domain reflectometer techniques. It sends a pulse down the cable. The pulse travels to the end and reflects at the impedance mismatch. The tester measures the time it takes to travel down the cable and back. The cable's velocity of propagation allows travel time to be equated to distance. The tester can calculate the distance based on the pulse's travel time and the cable's velocity of propagation.

One cable specification you might want to custom enter for a test is the cable's velocity of propagation. High-end testers offer a database of specific cables, such as AT&T 1061 or Belden 1588A. An error in the velocity of propagation will translate into an error in the calculated length. This is the same as calculating driving distance by knowing driving time. Velocity of propagation is the speed you drive. If you drive 60 miles an hour and a trip takes two hours, we can calculate you travelled 120 miles. If you drive 55 miles an hour for two hours, you have travelled 110 miles. If our database says velocity of propagation is 60 miles an hour, but your true speed is 55, then our calculations will be wrong by 10 miles. A tester will base velocity of propagation either on those recommended by the standard or by having specific cables in the database. If you have a different cable with a different velocity of propagation, you should enter it so that measurements will be more accurate. Remember that specifications for UTP cable, for example, are worst-case specifications. Many cables will have performance that will exceed the minimum requirements of the standard. In most cases, this simply gives you extra margin in meeting specifications. With velocity of propagation, the differences can lead to miscalculated lengths and distances.

A tester can calculate the nominal velocity of propagation on a known length of cable. If you are installing a type of cable not found in the database, you can determine the velocity of propagation on a cable 15 meters or longer.

Length measurements can also be used to locate breaks in the cable. A break will cause an end-of-cable reflection; the length reading will be obviously short, showing the distance to the break.

NEXT Loss

To measure NEXT, the tester injects a signal onto the transmit pair and records the energy on the receive pair.

A tester has multiple RF generators to generate radio-frequency signals over the bandwidth of interest, typically from 700 kHz to 100 MHz. The instrument sweeps the frequencies from low to high. Actually, it doesn't generate every frequency; rather, it works in user-selectable steps of

100, 200, or 500 kHz to generate over 1000 readings if readings are taken clear to 100 MHz. If you test Category 4 cable, the tester only generates frequencies to 20 MHz. For NEXT loss measurements, TSB-67 requires a maximum step size of 150 kHz for frequencies to 31.25 MHz and a step size of 250 kHz for frequencies above 31.25 MHz.

The tester also contains a tunable narrow-band receiver. Narrow band means that it only responds to a very narrow range of frequencies at a given time. A television or radio is a tunable narrow-band receiver. They tune to a station being received on a specific frequency range. They reject all other frequencies. The receiver tunes itself to the same frequency as the input signal. Extraneous noise from the surrounding environment – including strong radio signals, fluorescent lights, motors, and even noise from other cables – can be coupled onto the receive cable and added to the NEXT energy coupled from the transmitting or active line. A wide-band receiver would detect such noise and give invalid results, while a narrow-band receiver rejects it. The narrow-band receiver filters out any extraneous noise.

Attenuation

The attenuation measurement is straightforward. The signal injector launches a signal of known strength into one end of the cable, the tester measures the strength of the attenuated signal at the other end, and the attenuation is calculated in decibels. Results can also include a simple pass/fail indication. As with NEXT measurements, attenuation measurements are made over the entire frequency range.

Attenuation-to-Crosstalk Ratio

ACR is calculated from the NEXT and attenuation measurements. If problems occur with the ACR reading, it is best to review the NEXT and attenuation measurements and look for problems related to NEXT or attenuation.

Capacitance

Capacitance is the energy stored between the two conductors. For twisted pairs, capacitance is measured between the two wires of a pair. For coaxial cable, it is measured between the center conductor and shield. A given capacitance has an effective charging and discharging time constant. By measuring the time constant, capacitance can be calculated. Capacitance must fall into a medium range between two failing extremes.

DC Loop Resistance

DC loop resistance is the total resistance of both conductors of a pair or the center conductor and shield of a coaxial cable. A cable fails the loop resistance if its resistance exceeds the value allowed for the maximum cable length (most often, 100 meters). In terms of UTP, according to the 568 standard, a 50-meter cable fails only when its loop resistance exceeds that allowed for a 100-meter cable.

Impedance

A narrow impedance in the cable is important to reduce reflections and promote signal transmission without undue loss. The tester uses time-domain reflectometry to measure impedance at a point in the cable, typically between 25 and 40 meters away. The tester measures the reflected energy from that point. It's important to measure impedance at enough distance to go beyond any patch cables and cross connects. In a short cable, the impedance measurement will be the termination impedance.

While the impedance of a cable will vary along its length, this single measurement will tell whether the cable falls within the tolerances allowed. Category 3, 4, and 5 cables have a nominal impedance of 100 ohms and a tolerance of ±15 ohms. This means the allowed impedance can be anywhere from 85 ohms to 115 ohms. Most testers will display the actual measurement from each pair, as well as a pass/fail indication.

Pass/Fail and Accuracy

In making a pass/fail evaluation of the cable, the tester evaluates all pairs of the cable. If a single pair fails, the entire cable fails on that test. In other words, the tester evaluates on worst-case conditions. The tester will highlight the failure, indicating the pair that failed.

An important aspect of any measurement is its accuracy. A tester can have a margin of error of ±1.5 dB, meaning that the actual value being measured could be 1.5 dB on either side of the displayed value. The problem arises when the reading is very near the pass/fail point. NEXT loss for Category 5 cable is 41 dB at 25 MHz. Suppose the tester measures 42 dB and passes the cable at this frequency. Fine, we'll accept the evaluation, even though the real NEXT could be a failing 39.5 dB (within the 1.5 dB margin of error). Even worse, what if the tester gives a result of 40.5 dB and fails the cable? Since the measurement is within the margin of error, the nagging suspicion remains that the cable may, in fact, pass. You can retest, of course, but a failure remains a failure.

TSB-67 defines two levels of accuracy in field testers. Level II testers have a lower margin of error than Level I testers. For example, within tests required by TSB-67, here are the differences in accuracy:

	Level I	**Level II**
NEXT:	±3.4 dB	±1.6 dB
Attenuation:	±1.3 dB	±1.0 dB

Especially with NEXT loss, Level II testers are much more accurate than Level I testers.

TSB-67 requires that any measurement that falls within the uncertainty region of the tester's accuracy be flagged. Instead of reporting *Passed* or *Failed*, the tester will report **Passed* or **Failed*, with the asterisk indicating that the reading is close to the specification limit and within the uncertainty region. For the 41-dB limit for Category 5 cable at 25 MHz, a Level I tester will flag values between 37.6 and 44.4 dB, while a Level II tester will flag only those readings between 39.4 and 42.6 dB.

Reporting the Results

Each cable or link tested can be given an ID. A tester can store several hundred tests. The tester can then be connected to a computer to download the results into a data base or connected to a printer to print the results. Figure 8-4 shows a typical printout of a test. This printout forms part of the record used for administering the cable plant.

```
                        Godfrey Cyberwidgets
                 PENTASCANNER CABLE CERTIFICATION REPORT
                        CAT5 Link  Autotest

Circuit ID:         1                  Date:          20 March 95
Test Result:        PASS               Cable Type:    Cat 5 UTP
Owner:              PentaScanner       Gauge:
Serial Number:      38S95BB_KLW        Manufacturer:  Belden
Inj. Ser. Num:      38N94BB_KLW        Connector:     Mod Plug
SW Version:         V03.10             User:          DJS

Building:           Engineering        Floor:         Second
Closet:             2A-N
Rack:               1                  Hub:           10B-T/1
Slot:                                  Port:          13
```

Test		Expected Results		Actual Test Results			
Wire Map		Near: 12345678		Near: 12345678S			
		Far: 12345678		Far: 12345678S			
				Pr 12	Pr 36	Pr 45	Pr 78
Length	ft	10 -	328	95	95	97	95
Impedance	ohms	80 -	125	112	115	111	115
Resistance	ohms	0.0 -	18.8	4.8	5.2	4.9	5.1
Capacitance	pF	10 -	5600	1206	1189	1286	1219
Attenuation	dB			4.9	4.2	4.9	4.6
@Freq	MHz			100.0	100.0	100.0	100.0
Limit	dB			24.0	24.0	24.0	24.0

PENTA Pair Combinations		12/36	12/45	12/78	36/45	36/78	45/78
NEXT Loss	dB	39.3	40.0	37.5	32.8	37.6	39.5
Freq(0.7-100.0)	MHz	94.3	97.3	87.9	96.7	99.9	96.1
Limit: Cat 5 formula	dB	27.5	27.3	28.0	27.3	27.1	27.4
Active ACR	dB	38.3	45.6	34.9	32.1	33.0	36.1
Frequency	MHz	100.0	62.5	100.0	100.0	100.0	100.0
Limit: Derived	dB	3.1	12.1	3.1	3.1	3.1	3.1

INJ Pair Combinations		12/36	12/45	12/78	36/45	36/78	45/78
NEXT Loss	dB	37.0	38.6	34.2	31.7	38.9	43.6
Freq(0.7-100.0)	MHz	99.9	97.5	88.1	93.3	99.9	96.1
Limit: Cat 5 formula	dB	27.1	27.3	28.0	27.6	27.1	27.4
Active ACR	dB	32.1	38.8	36.4	31.9	34.3	38.9
Frequency	MHz	100.0	100.0	100.0	100.0	100.0	100.0
Limit: Derived	dB	3.1	3.1	3.1	3.1	3.1	3.1

```
Signature: _____     Date: _____
```

Figure 8-4. Printout of test results *(Courtesy of Microtest)*

TROUBLESHOOTING UTP CABLING

Some faults detected during testing and certification must be corrected. Some errors can easily be determined, others not so easily. The following gives some tips to help find the source of failure.

Testing can be done on the channel, link, and component level. If certification is done on a link level, a failure can occur on any cable or at interconnection in the path. To help target the point of failure, test individual cables in the link. This will help eliminate good cables and interconnections and highlight the bad one.

Wire Map Troubleshooting

Crossed pairs found by the wire map test are easily corrected by rewiring the connection. Be sure the tester is set for the wiring pattern being used. While all four pairs should be terminated, some applications-specific cabling may only wire the required pairs. Missing wires in the test indicates one of several conditions:

- Broken conductor
- Missed termination in the RJ45 or punch-down block
- Wrong test: for example, testing for a 568A wiring pattern when the connection is wired specifically for 10BASE-T or Token Ring. Or testing for 568B patterns when the wiring is 568A.

Length Fault Troubleshooting

Too long a cable is not usually a problem in a properly designed system. Check for coiled cable somewhere in the run or improper routing by an indirect path. In addition, length may not be a problem if the NEXT and attenuation limits are okay.

Length errors will occur if the cable's velocity of propagation differs from the value being used by the tester. Be careful if several different brands or types of cable are used. The velocity of propagation may have to be adjusted to handle different cables. Errors can also occur if you are testing a link having cables with different propagations of velocity.

The tester may inaccurately report a short length because of reflections that are punch-down blocks or patch panels. Normally, this is not a concern if all other tests are okay. Even so, you may want to check the connections.

Length measurements using TDR techniques can also locate opens and shorts. An open will cause a reflection indicating the end of a cable. A short will absorb the pulse so there is no reflection.

NEXT Troubleshooting

Next to wire map errors, NEXT problems are the main source of faults in the cabling system. NEXT errors derive from three sources: substandard components, installation errors, and too many interconnections in the path. Here are the main things to check:

- An old or improperly calibrated tester. If your tester was manufactured before 1995, check with the manufacturer about an upgrade to the latest version. There have been major improvements in test equipment recently and you should take advantage of this. Also, the testing requirements have been evolving quickly – TIA/EIA-568A and TSB-67 were only approved in mid-1995. Make sure even a recent tester has the latest firmware.

- The proper category of cable is used throughout.

- All other components – connectors, couplers, punch-down blocks, patch cables – are of the proper category.

- Incorrect patch cables. Double-check patch cables since they are the easiest to misapply. Make sure the category is correct. For any application, including low-speed voice, avoid the use of untwisted (silver satin) cables.

- Excessive untwist at the termination. Maintain the twist as close to the connector contacts as possible. This requirement is especially critical with Category 5 cable, where the untwist cannot exceed 0.5 inch.

- Too many interconnections. Reducing the number of interconnections in the path can reduce NEXT.

- Nearby noise sources. Even though the tester uses a narrow-band receiver to reduce its effects, ambient noise can affect readings. Power cables, fluorescent light, and so forth can couple energy onto the cable and be read as crosstalk. If necessary, reroute the cable away from noise sources.

Be sure to test NEXT at both ends of the cable. A cable or link can pass at one end and fail at the other.

Attenuation Troubleshooting

Attenuation errors almost always result from either using the wrong category of components or having a bad termination. Any point of termination increases resistance and hence attenuation. While a correctly terminated connection maintains an acceptably low resistance, the poor connection does not.

Attenuation-to-Crosstalk Ratio Troubleshooting

Since ACR measurements are based on NEXT and attenuation, the same troubleshooting tips apply. In particular check categories of components and the quality of terminations.

Capacitance Troubleshooting

Capacitance errors most often are caused by using a lower category of cable. Also check for broken conductors, shorts, opens, and excessive noise.

Loop Resistance Troubleshooting

Too high a loop resistance typically has two causes:

- Excessive cable length. Check the cable length.

- Poor terminations. An improper termination adds significant resistance.

Impedance Troubleshooting

Impedance errors should not occur if the proper cable has been installed. Impedance errors most often occur if a substandard cable is installed. Notice that the Categories 2, 3, 4, and 5 cables all share the same impedance value, so impedance measurements will not indicate the wrong category. Measurements showing improper impedance will also detect other problems that should be investigated.

While 100-ohm UTP is the overwhelming favorite for premises cabling, remember that ISO/IEC 11801 also recognizes a 120-ohm cable. Make sure this cable isn't being used if you experience impedance errors.

Recurring Problems

It's quite common to find a few errors to correct during certification. The most common are wiring errors at interconnections, either crossed and split pairs or poor workmanship. But what do you do if large parts of the installation fail?

First, make sure the tester is calibrated. Next, make sure you are using the correct suite of tests, testing apples to apples and not to oranges.

Most often, these confusions arise from misunderstandings over what is being tested. Link testing, for example, is not a mandatory part of TIA/EIA-568A. Many installations are to be tested to 568, but without specifying exactly what is meant. For some, this means link testing; for others, it means cable testing. There are lots of numbers flying around – specs for cables, connectors, and links – so that it's easy to compare one test to another set of results. A link test will always fail if only the horizontal cable specs are used. Link specifications are about 20% looser than component specifications. Always make sure the tester is actually set for the test being performed.

TSB-67 addresses these problems by providing clear definitions of link and channels, performance specifications for each, testing requirements, and accuracy levels. These results should allow much less confusion about what's being tested.

FIBER TESTING

Fiber testing is simpler than UTP testing in that fewer characteristics are measured. In fact, testing can be done simply by measuring power with an optical power meter. The key to any fiber-optic link is the power budget: loss of optical power through the link cannot exceed specified limits. Power loss is the main measurement made.

Certification is done on link level. Begin with a basic link test. If the link passes, component-level testing is not necessary. If the link fails, test individual components to isolate the fault.

Continuity Testing

Simple continuity testing can be achieved by a flashlight: does the light come through the fiber? Fancy flashlights – called visual continuity testers – are available specifically for fiber-optic testing. A red light is preferred since it is easiest to see.

Warning: Never Look into an Energized Fiber

Never look into a fiber that might be energized. Fiber-optic systems use infrared light that cannot be seen. Laser and some LED sources emit light powerful enough to damage the eye. Before looking into a fiber, either make sure the system is powered down or that the fiber is disconnected from any transmitters.

Link Attenuation

Link attenuation is simply the sum of the individual losses from the cable, connectors, and splices. Cable loss can be assumed to be linear with distance. For the 90-meter horizontal cable run, a 62.5/125-μm cable has an attenuation of 0.34 dB at 850 nm and 0.14 dB at 1300 nm. Maximum insertion loss values are 0.75 dB for connectors and 0.3 dB for splices. To estimate the total loss in the link, simply add up the individual losses. Consider a horizontal run with two connectors. The total loss is 1.84 dB:

Cable loss:	3.75 dB/km X 90 m	= 0.34 dB
Connectors:	0.75 dB X 2	= 1.5 dB
Total loss:		= 1.84 dB

TIA/EIA recommends a maximum attenuation for the 90-meter horizontal cable run of 2.0 dB, including the cable and two connectors (one in the telecommunications closet and one in the work area). If passive links are connected serially, the expected loss is the sum of the individual links. Notice that an interconnection includes the connectors on both sides of an adapter. Be careful not to count an interconnection twice. Consider the work area outlet. If you counted this interconnection as part of the horizontal cable, don't count it as part of the work area patch cable.

Notice, too, that TIA/EIA-568A link performance does not include passive elements like couplers, switch, and other optical devices. These can add appreciable loss in the link and must be accounted for if present. Links using such devices will not meet 568 requirements and must be evaluated on an individual basis. For example, 10BASE-FP networks use a passive star coupler to divide light among many fibers. This system would have to be evaluated in terms of IEEE 802.3 10BASE-F specifications rather than EIA-TIA-568A specifications.

In estimating cables losses, use the following values:

62.5/125-μm cable at 850 nm:	3.75 dB
62.5/125-μm cable at 1300 nm:	1.5 dB
Single-mode outside plant cable at 1310 nm:	0.5 dB
Single-mode outside plant cable at 1550 nm:	0.5 dB
Single-mode inside plant cable at 1310 nm:	1.0 dB
Single-mode inside plant cable at 1550 nm:	1.0 dB

Measured loss must be less than those calculated by worst-case values. However, some care in testing is necessary because the test procedure can significantly affect results.

UNDERSTANDING FIBER OPTIC TESTING

A fiber carries light in modes (possible paths for the light). The modal conditions of light propagation can vary, depending on many factors. We generally speak of three general conditions: overfilled fiber, underfilled fiber, and equilibrium mode distribution.

- **Overfilled fiber.** When light is first injected into a fiber, it can be carried in the cladding and in high-order modes. Over distance, these modes will lose energy. The cladding, for example, attenuates the energy quickly. High-order modes can be converted into lower order modes. Imperfections in the fiber can change the angle of reflection or the refraction profile of high-order modes.

- **Underfilled fiber.** In some cases, light injected into the fiber fills only the lowest order modes. This most often happens when a low NA laser is used to couple light into a multimode fiber. Over distance, some of this energy shifts into higher modes.

- **Equilibrium mode distribution.** For both overfilled and underfilled fibers, optical energy can shift between modes until a steady state is achieved over distance. This state is termed *equilibrium mode distribution* or *EMD*. Once a fiber has reached EMD, very little additional transference of power between modes occurs.

EMD has several important consequences in light propagation in a fiber and its effects on testing. Both the active light-carrying diameter of the fiber and the NA are reduced. For example, the fiber recommended for premises cabling has a core diameter of 62.5 μm and an NA of 0.275. These are based on physical properties of the fiber. With EMD, we are interested in the *effective* diameter and NA. In an 62.5-μm fiber at EMD, the effective (light-carrying) core diameter is 50 μm and the NA is less than 0.275.

So what's the point in being concerned with EMD? The modal conditions of the fiber affect the loss measured at a fiber-to-fiber interconnection. It is easier to couple light from a smaller diameter and NA into a larger diameter and NA. Even so, the light coupling into the second fiber will not be at EMD. It will overfill the fiber to some degree so that EMD will not be achieved until the light has travelled some distance.

> *In testing, it is important to specify the modal conditions. Testing a connector at EMD can yield significantly different results than testing it under overfilled conditions. A connector can yield an insertion loss of 0.3 dB under EMD conditions and 0.6 dB under fully filled conditions.*

In testing, we speak of launch and receive conditions:

Long launch:	Fiber is at EMD at the end of the launch fiber.
Short launch:	Fiber is overfilled at the end of the launch fiber.
Long receive:	Fiber is at EMD at the end of the receive fiber.
Short receive:	Fiber is overfilled at the end of the receive fiber.

EMD conditions can be easily simulated without a long length of fiber. The standard method of achieving EMD in a short length of fiber is by wrapping the fiber five times around a mandrel. This mixes the modes to simulate EMD.

STANDARD TESTS

There are three standard tests applicable to premises cabling.

OFSTP-14:	link certification with power meters
FOTP-171:	patch cable certification
FOTP-61:	link certification with an optical time-domain reflectometer

Let's begin by looking at power meters.

Power Meters

An optical power meter does just that: measures optical power. Light is injected into one end and measured at the other. Some meters incorporate interchangeable light sources into the meter so that it serves as either a light source or a meter (or both simultaneously). Most meters can display power levels in user selectable units, most often dB or dBm. The dB reading is most appropriate for certification. Because fiber performance depends significantly on the wavelengths propagated, the light source is important. TIA/EIA-568A makes the following recommendations for wavelengths and testing:

Horizontal multimode cable:	either 850 or 1300 nm
Backbone multimode cable:	both 850 and 1300 nm
Backbone single-mode cable:	both 1310 and 1550 nm

A power meter typically requires two readings. One is used for calibration, to zero the meter. The other is the actual power reading. Of course, calibration isn't required for each reading; once the unit is calibrated, it can be used for numerous tests. Check the meter's user's guide for direction on how often to zero the unit.

An optical tester similar to a power meter is the optical loss test set (OLTS). The OLTS is distinguished from the power meter by including the signal injector in the same unit as the power meter.

Link Certification with a Power Meter (OFSTP-14)

OFSTP-14 offers two measurement procedures, Method A and B.
Both methods are similar:

1. A calibration reading is taken using one or two test jumpers.

2. The jumpers are then connected to the link.

3. The link loss is displayed.

Method A uses the two test jumpers for the calibration, while Method B uses only one. Both methods use two jumpers for the link evaluation. Why the difference? The single-jumper approach of Method B eliminates the influence of the connectors used to interconnect the two cables during the zeroing procedure. The connectors in the middle of the two jumpers essentially mirror the presence of the connectors in the actual link. Test Method A cancels the influence of these connectors during the test. Method B includes these connectors in the test.

Of the two methods, TIA/EIA-568A recommends Method B. Figure 8-5 shows the general setup and sequence for the test. Keep in mind that the figure is general; it does not show patch panels and other intervening couplings that could be included in the test.

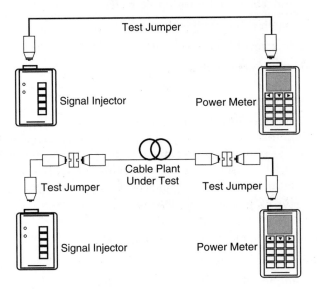

Figure 8-5. Link certification with a power meter

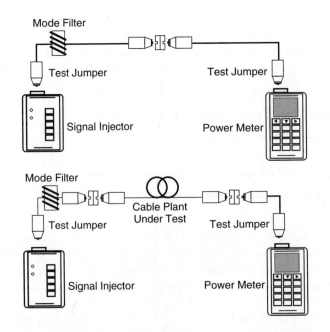

Figure 8-6. Measuring cable attenuation with a power meter

Testing Patch Cables (FOTP-171)

This procedure includes four methods, with three optional procedures for each method. We will restrict our discussion to Method B, which is the best suited to premises cabling and is the method typically used by cable assemblers.

This procedure (Figure 8-6) uses the substitution method to evaluate attenuation in a length of fiber. Power though two jumper cables are measured to zero the meter. The cable assembly to be tested is then inserted between the two jumper fibers. The new reading is the fiber attenuation. Notice that the test requires a mode filter on the launch end to simulate EMD in the short length of jumper cable. This can be achieved by wrapping the cable five turns around a mandrel to mix the modes.

Optical Time-Domain Reflectometry

An optical time-domain reflectometer (Figure 8-7) is a useful tool for troubleshooting. We discussed the uses of time-domain reflectometry in testing UTP links. Certification tools use TDRs to measure cable lengths. OTDR – optical TDR – is more widely used than metallic TDR. The principles are the same: a pulse is sent down the fiber. Energy reflecting from imperfections and changes in refractive index of the path are detected and displayed. Changes in the index of refraction of a fiber are analogous to changes in impedance in a copper cable: both cause reflections that can be used to get a clear picture of the link.

Most TDRs are highly automated and sophisticated, allowing you to zoom into specific areas for a close inspection. Among capabilities are these:

- Zoom in to specific events like connectors, splices, and faults (all of which will show pronounced bumps in the trace) for a closeup look.

- Zoom out to gain an overall picture of the link.

Figure 8-7. Optical time-domain reflectometer *(Courtesy of Photon Kinetics)*

- Measure the total link length.
- Measure distances to specific spots.
- Measure attenuation for the entire link or for specific areas.
- Compare two sections of a link. You can, for example, compare two different interconnections.
- Compare two different links. Many OTDRs have storage capability that will allow you to store a test. This test can then be called up and compared to the current test.
- Save the test results to a computer database.
- Print the test results.

Do you need an OTDR? While link certification can be done with a power meter, nothing beats an OTDR for in-depth troubleshooting. The OTDR has two main drawbacks. First, they are relatively expensive. Second, they require a higher level of operator skill. Even so, OTDRs in recent years have become lower in cost and easier to use. Newer handheld models replace the bulkier "lugarounds" that usually required a cart.

A useful strategy is to perform certification using power meters, but have one or two OTDRs available for troubleshooting. This can additionally save on training, since only one or two technicians need to specialize in OTDR use.

Some building owners or network administrators, however, require an OTDR plot for every fiber-optic link. This is a piece-of-mind issue, not a performance issue. Compared to a UTP link report – detailing NEXT, attenuation, ACR, loop resistance, capacitance, and wire mapping – an optical link report seems short and insubstantial. An OTDR plot adds weight.

Measuring Link Attenuation with OTDR (FOTP-61)

Figure 8-8 shows the general setup for measuring link attenuation with an OTDR. A dead-zone fiber is required at each end of the link. An OTDR cannot "see" the fiber close to it. In order to overcome this deficiency, a short length of fiber is inserted between the instrument and the cable being tested. How long the dead-zone fiber must be depends on the specific OTDR being used. OTDRs designed for long-distance telecommunications testing typically have a longer deadzone than units more appropriate for premises cabling. The second dead-zone fiber is required to allow you to evaluate the final connector at the far end. when the OTDR pulse reaches the end of the fiber, it reflects a large amount of energy. If you end on the cable under test, then the reflected pulse will include no information about the quality of that connection.

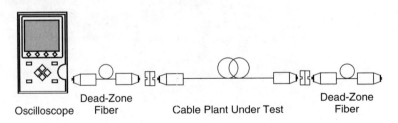

Oscilloscope   Dead-Zone Fiber   Cable Plant Under Test   Dead-Zone Fiber

Figure 8-8. Measuring link attenuation with an OTDR

TROUBLESHOOTING AN OPTICAL LINK

This section discusses some of the things to check if the link fails.

- Continuity. Use a visual continuity tester – a flashlight – to perform a continuity check. This will help identify broken fibers.

- Polarity. Were the fibers crossed during termination? Check end-to-end continuity for each section of the link to discover crossed pairs. Reterminate the cable, or reconnect the cable properly.

- Misapplied connector. Check the connector carefully. Check the end finish with a microscope for unacceptable scratches, pits, or other flaws. As appropriate, repolish the connector or reterminate the fiber.

- Dirt. Check the end of the fiber for dirt or films. Clean the connector end with isopropyl alcohol or canned air. Check the core of the coupling adapter for obstructions.

- Make sure the light source and power meter are correctly set and calibrated. For example, using a 850-nm source to test against 1300-nm test conditions significantly distort results.

- Check the cable plant for tight bends, kinks in the cable, and other unhealthy conditions. These can either increase attenuation or snap fibers in two.

- Make a habit of checking the cable for continuity before installing it. Test it while it is still on the reel. Considering the time it takes to perform this test versus the cost of installing the cable, it is an inexpensive first step in ensuring a quality installation.

A NOTE ON ISO/IEC TESTING REQUIREMENTS

ISO distinguishes three types of testing and recommends the types of tests "likely" to be performed on each. Thus, the tests are not absolutely required by the standard, but any testing done should be inclusive enough to ensure the performance of the cabling system. The three types of testing identified are:

- Acceptance Testing, which is used to validate installed cabling that is known to meet the requirements for cable types, connecting hardware, and proper topology and distance. In other words, if you have installed Category 5 cable and hardware and met the distance requirements, acceptance testing is all that is needed to test a link. Notice that all this level requires is a simple wire map test for UTP and STP cable. This approach, however, will not guarantee performance if poor workmanship or other installation error has occurred. Shoddy workmanship can reduce Category 5 components to a Category 3 link.

- Troubleshooting, which is a series of tests that can aid troubleshooting of components and links. The standard suggests you choose the test most appropriate to solving the problem.

- Compliance Testing, which is used to ensure that the cabling plant meet the link performance requirements. This series of test can be used on cabling systems of known or unknown components. Thus, if you don't know what categories of components were originally used in the cabling plant, these tests can help determine the level.

Figure 8-9 lists the tests suggested for each type of testing.

Cabling System	Test	Acceptance	Troubleshooting	Compliance
UTP/STP	Characteristic Impedance			✔
	Propagation Delay			✔
	DC Resistance			✔
	NEXT Loss		✔	✔
	Attenuation		✔	✔
	Return Loss		✔	✔
	Shield DC Resistance		✔	
	Echo Response		✔	
	Wire Map (Continuity)	✔	✔	
Optical	Multimode Modal Bandwidth			✔
	Propagation Delay		✔	✔
	Attenuation	✔	✔	✔
	Return Loss	✔	✔	✔

Figure 8-9. ISO/IEC 11801 Link Testing Recommendations

CHAPTER 9

Documenting and Administrating the Installation

The cabling structure is not static. It is dynamic, growing and evolving with the changing needs of users. Maintaining up-to-date, reliable documentation of the cabling system is essential. Without such documentation, the wiring system descends into chaos until nobody knows how the system is arranged.

Careful, accurate administration of the cabling system has several benefits:

- To allow fast, accurate moves, adds, and changes.
- To speed and simplify troubleshooting.
- To increase network reliability and uptime.
- To allow better asset management.
- To increase both management and MIS people's confidence in structured cabling systems. Some MIS people, in particular, may question whether a structured cabling system offers the same reliability as a proprietary, fixed system. As businesses "downsize" from mainframe and minicomputer-based systems to LAN-based systems, this question is significant to those responsible for the computer infrastructure.
- To allow disaster recovery plans for the cable plant to be developed.
- To generate management reports.
- To allow reliable capacity planning to manage the needs from growth and migration in both the cabling plant and in executed applications. In other words, careful administration will help prove (or disprove) the ability of the plant to accommodate emerging applications like ATM.
- To allow new tenants to receive accurate, up-to-date information on the cabling plant. Such information can enhance the value of a building.

TIA/EIA-606 is a standard entitled *Administration Standard for the Telecommunications Infrastructure of Commercial Buildings*. It provides guidance on documenting a cabling system to make its administration efficient and effective.

WHAT'S IN THE DOCUMENTATION?

Documentation includes labels, records, drawings, work orders and reports.

Labels

Every cable, piece of termination hardware, cross connect, path panel, conduit, closet, and so forth should be labeled with a unique identifier. While the label can simply be a number, it's wiser to create a code system that will provide meaningful information. For example, you could label the cable "C12345." No other cabling component will have this number, but the label tells you nothing about the cable. A better approach would be to label the cable "TC3A-18-WA16." This means cable 18, running from telecommunications closet 3A to work area 16.

Records

No matter how good your coding scheme for components, you cannot put all the information required for decent documentation on the label. The label also serves as a link to a record that contains complete information on the component. The record can exist either on paper or in a computer file. If you type "TC3A-18-WA16" into the computer database, up will pop the complete record for that cable.

TIA/EIA-606 breaks a record into four types of information:

1. Required information. This is the essential information about the component. For a cable, it includes the type (UTP Category 5, 62.5/125 fiber, etc.), unterminated pairs, damaged pairs, and available pairs).

2. Required linkages. Linkages are pointers to other records. For a cable, it includes termination records showing where and how the cable is terminated, and pathway records showing where and how the cable is routed.

3. Optional information. This includes any information you want to include to make the record as complete and comprehensive as you require. It can include the cable's length, product code or manufacturer's part number, test information, and so forth.

4. Optional linkages. This points to additional records that might be helpful to include. For example, if a cable attaches to a network hub or workstation, you might want to include a linkage to the record for the device.

Drawings

Drawings are usually based on the architectural drawings for the building. They help locate components within the building. It's of little help to know that a cable terminates in closet 3A, if you don't know where the closet is. Drawings also show the locations of conduits, pull boxes, and other components hidden from view behind walls, above ceilings, and under floors.

Work Orders

Work orders record all moves, adds, and changes. They form a history of the cabling system's life. An equally important practice is to update the records each time a work order is executed.

Reports

A report is a group of records organized in a specific manner. A computer database for a cabling system can allow you to generate specific reports of selected parts of records. An example is a cable report showing all cables running from a wiring closet. The report can include the cable identifier, pathway, termination position in the closet, termination position in the work area, cable length, and application.

WHAT'S IN A RECORD?

A record can be as comprehensive as you want. TIA/EIA-606 makes specific recommendations regarding required and optional information and linkages. Figure 9-1 summarizes these requirements.

COLOR CODING

Color coding of circuits can be useful for identifying circuits by type or function. Some confusion exists regarding stands-based color coding. TIA/EIA-606 recommends color coding only for cross connects within the cabling system. Figure 9-2 summarizes the color coding. Proper color coding is a valuable aid in making moves, adds, and changes and in troubleshooting the cabling plant. Color coding allows technicians to identify cables quickly and to double-check work readily. Color coding reduces wiring errors.

Your color-coding scheme does not have to be limited to those recommended in 606. Some users also color code by level of performance so that a Category 3 connection is easily distinguished from a Category 5 connection. While 606's color coding is intended for cross connects, it can be extended to other parts of the building. For example, the blue recommended for horizontal cabling is for the telecommunications closet, not for the outlet in the work area. Color-coding an outlet, however, for telephones and low- and high-speed network connections can help identify ports as well as icons and words.

CABLE MANAGEMENT SOFTWARE

While a basic computer database can be used to maintain records and generate reports, programs are also available specifically for managing a cabling system. The programs are highly compatible with structured cabling systems and standards like TIA/EIA-568A and TIA/EIA-606.

Programs range from simple, inexpensive spreadsheets and databases to elaborate, expensive graphical programs. The simplest program can be created by people with modest skills in adapting spreadsheets and databases from general-purpose programs. The more sophisticated one, especially those with advanced graphics, are specialized packages. Likewise, some programs require modest computer resources to run, while others take a computer with considerable horsepower and disk space to make their use practical.

Cable	Required Information	Cable Identifier
		Cable Type
		Unterminated Pairs/Conductor Numbers
		Damaged Pairs/Conductor Numbers
		Available Pairs/Conductor Numbers
	Required Linkages to Other Records	Termination Position
		Splice
		Pathway
		Grounding
Termination Hardware	Required Information	Termination Hardware Identifier
		Termination Hardware Type
		Damaged Position Numbers
	Required Linkages to Other Records	Termination Position
		Space
		Grounding
Termination Position	Required Information	Termination Position Identifier
		Termination Position Type
		User Code
		Cable Pair/Conductor Numbers
	Required Linkages to Other Records	Cable
		Other Termination Positions
		Termination Hardware
		Space
Splice	Required Information	Splice Identifier
		Splice Type
	Required Linkages to Other Records	Cable
		Space
Pathway	Required Information	Pathway Identifier
		Pathway Type
		Pathway Fill
		Pathway Loading

Figure 9-1. TIA/EIA record summary: required information and linkages

Figure 9-1 Continued

	Required Linkages to Other Records	Cable
		Space
		Pathway
		Grounding
Space	Required Information	Space Identifier
		Space Type
	Required Linkages to Other Records	Pathway
		Cable
		Grounding
Telecommunications Main Grounding Busbar	Required Information	TMGB Identifier
		Busbar Type
		Grounding Conductor Identifier
		Resistance to Earth
		Date Measurement Taken
	Required Linkages to Other Records	Bonding Conductor
		Space
Bonding Conductor	Required Information	Bonding Conductor Identifier
		Conductor Type
		Busbar Identifier
	Required Linkages to Other Records	Grounding Busbar
		Pathway
Telecommunications Grounding Busbar	Required Information	Busbar Identifier
		Busbar Type
	Required Linkages to Other Records	Bonding Conductor
		Space

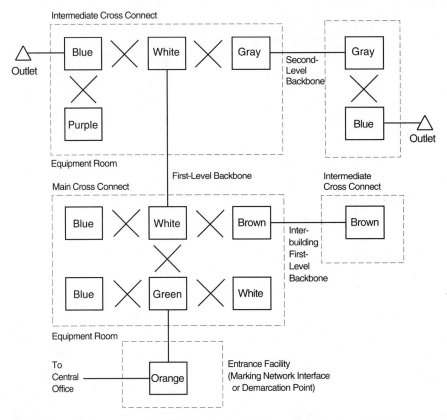

Orange	Demarcation point:	termination of cable from telephone company's central office
Green	Networks:	network connections or auxiliary circuit terminations
Purple	Common Equipment:	switching and data equipment (PBXs, hubs, etc.)
White	First-level backbone:	connections between main and intermediate cross connects
Gray	Second-level backbone:	connections between intermediate and horizontal cross connects
Blue	Station:	horizontal cable terminations
Brown	Interbuilding backbone	
Yellow	Miscellaneous circuits:	alarms, security, environmental controls, etc.
Red	Key telephone systems	

Figure 9.2 Cross connect color coding

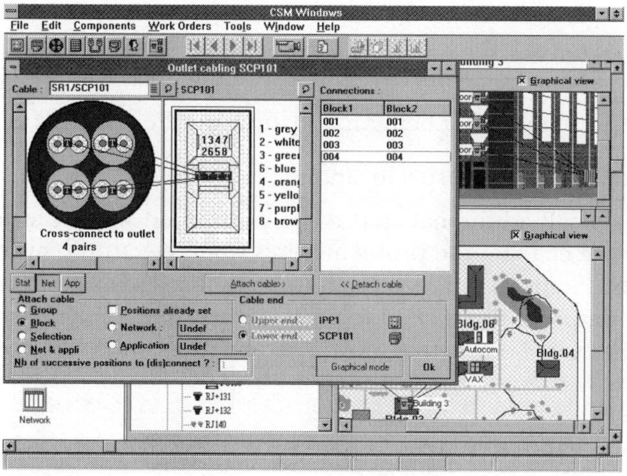

Figure 9-3. Cable management software eases documentation and administration *(Courtesy of Apsylog)*

The cable management program shown in Figure 9-3 uses a graphical interface to allow both physical and logical views of the cabling system. The graphical interface helps you to visualize and understand the cabling system. By providing the ability to view the system at multiple levels, from an overall building-to-building layout down to wiring of cross connects, patch panels, and work area outlets, the program makes it simple to design and review the cabling system. A built-in library of components – cross connects, patch panels, hubs, cables, and so forth – makes it fairly easy and straightforward to create your system graphically.

While simply viewing the cabling plant on multiple levels might be beneficial, the power of cable management systems is in helping you maintain the system. The program can automatically validate data paths pair by pair or for an entire subsystem. This doesn't mean the program can identify installation errors where the cable was not run according to the plan. Cable management programs do not connect directly to the wiring, so they can only validate that the plan is correct, or that the cabling is "theoretically" interconnected properly. Installers must still physically verify the installation with testers. Still, the ability to logically check the cabling plant quickly and automatically beats manual checking.

Programs should also be able to produce detailed work orders based on proposed changes in the cabling. Once the work order has been carried out, the program will modify the database to reflect the changes.

Moves, Adds, and Changes

As part of the life cycle of the cabling plant, moves, adds, and changes must be accommodated by any administrative system. Failure to document all changes and update all related records in a timely, structured manner can result, over time, in a cabling plant whose physical structure no longer matches the record and drawings that supposedly document it. Inaccuracies increase the likelihood of errors occurring during moves, adds, and changes. Network reliability degrades, downtime increases, and users are peeved.

Should you recertify after each move, add, or change? Testing each link segment affected by a move, add, or change is a good idea. The reassurance of knowing that the link is still within spec is a quick indication that no errors were made. Retesting has two principle benefits:

- It updates any performance data being maintained on the cable plant.

- It ensures that the link is within specification.

While testing is a small additional step to be performed while the technician is in the wiring closet making the change, the problem arises when electronic switches allow automatic switching of circuits. (Some newer switches have built-in measurement capabilities.) Since a technician no longer has to be dispatched to perform the change, there is less impetus to test. Still, testing is the only way to be sure that moves, adds, and changes don't bring the cable plant out of spec.

ELECTRONIC ADMINISTRATION

In Chapter 4, we described an electronic switch – a form of patch panel where circuits are connected internally to the device rather than manually. Electronic switches permit easy moves, adds, and changes from a central management computer. The administrator simply defines how the wiring should be re-arranged, the order is issued, and the change is accomplished. For the administrator, the benefits are obvious. There is no walking to the equipment closet, no manually plugging and unplugging circuits, no danger of making a poor connection in the rearrangement, and less paperwork to manage. Move, adds, and changes can be done in minutes.

The drawback to the electronic switch is the possibility of failure. If the switch fails catastrophically, the entire cabling system – or at least that part attached to the switch – is brought down. Manual cross connects seldom fail on their own.

CONVERGENCE OF LANS AND CABLING SYSTEMS

Networks are managed by software (generally called a network management system or NMS) that monitors their operation. The software can alert administrators of faults or impending faults, allow stations to be added or removed from the network, generate automatic maps of how the network is connected, generate statistical reports to allow planning, and permit network devices to be configured to varying degrees. By far the most popular set of rules for managing network devices is the Simple Network Management Protocol (SNMP).

While most PBXs traditionally use proprietary management, many vendors are committed to offering an SNMP agent to allow them to be managed by an NMS. Moves, adds, and changes can be accomplished by the network administrator from within the network management software rather than through a separate program. As structured cabling systems become more popular, you can expect tighter integration of network management systems, cable management software, and switching devices.

Appendix A
PREMISES CABLING SPECIFICATIONS

This appendix summarizes the specifications of most interest to those who test and certify premises cabling. It does not include other characteristics of concern to manufacturers in designing and categorizing components.

TIA/EIA-568A AND TSB-67

UTP Cables

Characteristic	Frequency	Category 3	Category 4	Category 5
NEXT Loss	0.772	43	58	64
	1	41	56	62
	4	32	47	53
	8	27	42	48
	10	26	41	47
	16	23	38	44
	20	–	36	42
	25	–	–	41
	31.25	–	–	39
	62.5	–	–	35
	100	–	–	32
Attenuation	0.772	2.2	1.9	1.8
	1	2.6	2.2	2.0
	4	5.6	4.3	4.1
	8	8.5	6.2	5.8
	10	9.7	6.9	6.5
	16	13.1	8.9	8.2
	20	–	10.0	9.3
	25	–	–	10.4
	31.25	–	–	11.7
	62.5	–	–	17.0
	100	–	–	22.0
Impedance		100 ohms ±15%		
Structural Return Loss	1	12	21	23
	4	12	21	23
	10	12	21	23
	16	10	18	23
	20	–	19	23
	31.25	–	–	20.9
	62.5	–	–	18
	100	–	–	16
Mutual Capacitance (pF)	1 kHz	6.6	5.6	5.6

NEXT Loss

TIA/EIA-568A supplies a formula for calculating NEXT loss at any frequency: NEXT loss at any given frequency is based on the NEXT loss requirement of 0.772 MHz (772 kHz). The 0.772-MHz values are given in the table on the previous page.

$$\text{NEXT}(f) \geq \text{NEXT}(0.772) - 15 \log (f / 0.772)$$

where f is the frequency in megahertz.

Attenuation

Attenuation in TIA/EIA-568A for UTP cables is based on the following:

$$\text{Attenuation } (f) \leq k1 \sqrt{f} + k2\, f + k3/ \sqrt{f}$$

where k1, k2, and k3 are constants:

	k1	k2	k3
Category 3	2.320	0.238	0
Category 4	2.050	0.043	0.057
Category 5	1.967	0.023	0.050

Structural Return Loss

Structural return loss is based on the following:

Frequency	Category 3	Category 4	Category 5
1–10 MHz	12	21	23
10–16 MHz	$12 - 10 \log (f/10)$	$21 - 10 \log (f/10)$	23
16–20 MHz	—	$21 - 10 \log (f/10)$	23
20–100 MHz	—	—	$23 - 10 \log (f/20)$

UTP Connecting Hardware (TIA/EIA-568A)

	Frequency	Category 3	Category 4	Category 5
NEXT Loss	1	58	65	65
	4	46	58	65
	8	40	52	62
	10	38	50	60
	16	43	46	56
	20	–	44	54
	25	–	–	52
	31.25	–	–	50
	62.5	–	–	44
	100	–	–	40
Attenuation	1	0.4	0.1	0.1
	4	0.4	0.1	0.1
	8	0.4	0.1	0.1
	10	0.4	0.1	0.1
	16	0.4	0.2	0.2
	20	–	0.2	0.2
	25	–	–	0.2
	31.25	–	–	0.2
	62.5	–	–	0.3
	100	–	–	0.4
Return Loss	1-20	–	23	23
	20-100	–	–	14

Connecting Hardware NEXT Loss (TIA/EIA-568A)

TIA/EIA-568A supplies a formula for calculating NEXT loss of connecting hardware at any frequency: NEXT loss at any given frequency is based on the NEXT loss requirement of 16 MHz. The 16-MHz values are given in the table above.

$$\text{NEXT}(f) \geq \text{NEXT}(16) - 20 \log (f/16)$$

Channel Limits (TSB-67 and TIA/EIA-568A, Annex E)

Characteristic	Frequency	Category 1	Category 4	Category 5
NEXT Loss	1	39.1	53.3	60.0
	4	29.3	43.3	50.6
	8	24.3	38.2	45.6
	10	22.7	36.6	44.0
	16	19.3	33.1	40.6
	20	–	31.4	39.0
	25	–	–	37.4
	31.25	–	–	35.7
	62.5	–	–	30.6
	100	–	–	27.1
Attenuation	1	4.2	2.6	2.5
	4	7.3	4.8	4.5
	8	10.2	6.7	6.3
	10	11.5	7.5	7.0
	16	14.9	9.9	9.2
	20	–	11.0	10.3
	25	–	–	11.4
	31.25	–	–	12.8
	62.5	–	–	18.5
	100	–	–	24.0

10:15
...ortland to ~~Roseberg~~
Eugene

Depart
15 Bus from Train Amtrack
12:15 · 2:40 11.58
5:07

Depart   Arival
10:40   5:15

$ 30

...cy	Category 1	Category 4	Category 5
	40.1	54.7	60.0
	30.7	45.1	51.8
	25.9	40.2	47.1
	24.3	38.6	45.5
	21.0	35.3	42.3
	–	33.7	40.7
	–	–	39.1
	–	–	37.6
	–	–	32.7
	–	–	29.3
	3.2	2.2	2.1
	6.1	4.3	4.0
	8.8	6.0	5.7
	10.0	6.8	6.3
	13.2	8.8	8.2
	–	9.9	9.2
	–	–	10.3
	–	–	11.5
62.5	–	–	16.7
100	–	–	21.6

NEXT Loss

Link NEXT loss is the vector sum of crosstalk induced in the cable, connectors, and patch cables. It is given by

$$\text{NEXT}(f) = -20 \log \sum 10^{-N_i/20}, i = 1, 2, \ldots n$$

where N_i is the crosstalk of component i at frequency f and n is the number of components at the near end. The near end is up to 20 meters. However, NEXT should be measured at both ends of the link.

Optical Fibers

Fiber-Optic Cable

	Multimode Fiber		Single-Mode Fiber
	850 nm	**1300 nm**	**1310 nm and 1550 nm**
Bandwidth (MHz-km)	160	500	—
Attenuation (dB/km)	3.75	1.5	1.0 (indoor cable) 0.5 (outdoor cable)

Fiber-Optic Connecting Hardware

Connector Insertion Loss (dB)		0.75 maximum
Splice Insertion Loss (dB)		0.3
Durability		500 mating cycles
Return Loss (dB)	Multimode	20
	Single Mode	26
Preferred Connector		SC

ISO/IEC 11801

UTP

Characteristic	Frequency	Category 3		Category 4		Category 5	
NEXT Loss	0.772	43		58		64	
	1	41		56		62	
	4	32		47		53	
	10	26		41		47	
	16	23		38		44	
	20	–		36		42	
	31.25	–		–		40	
	62.5	–		–		35	
	100	–		–		32	
Attenuation		100 Ω	120 Ω	100 Ω	120 Ω	100 Ω	120 Ω
	0.772	2.2		1.9	1.7	1.8	1.5
	1	2.6		2.1	2.0	2.1	1.8
	4	5.6		4.3	4.0	4.3	3.6
	10	9.8		7.2	6.7	6.6	5.2
	16	13.1		8.9	8.1	8.2	6.2
	20	–		10.2	9.2	9.2	7.0
	31.25	–		–		11.8	8.8
	62.5	–		–		17.1	12.5
	100	–		–		22.0	17.0
Impedance		100 ohms ±15% or 120 ohms ± 15%					
Structural Return Loss	1	12		21		23	
	4	12		21		23	
	10	12		21		23	
	16	10		19		23	
	20	–		18		23	
	31.25	–		–		20.9	
	62.5	–		–		18	
	100	–		–		16	
Mutual Capacitance (pF)	1 kHz	6.6		5.6		5.6	

UTP Connecting Hardware

	Frequency	Category 3	Category 4	Category 5
NEXT Loss	1	58	>65	>65
	4	46	58	>65
	10	38	50	60
	16	34	46	56
	20	–	44	54
	31.25	–	–	50
	62.5	–	–	44
	100	–	–	40
Attenuation	1	0.4	0.1	0.1
	4	0.4	0.1	0.1
	10	0.4	0.1	0.1
	16	0.4	0.2	0.2
	20	–	0.2	0.2
	31.25	–	–	0.2
	62.5	–	–	0.3
	100	–	–	0.4
Return Loss	1-20	–	23	23
	20-100	–	–	14

Link Classes

	Frequency	Class A	Class B	Class C	Class D
NEXT Loss	0.1	27	40	–	–
	1	–	25	39	54
	4	–	–	29	45
	10	–	–	23	39
	16	–	–	19	36
	20	–	–	–	35
	31.25	–	–	–	32
	62.5	–	–	–	27
	100	–	–	–	24
Attenuation	0.1	16	5.5	–	–
	1	–	5.8	3.7	2.5
	4	–	–	6.6	4.8
	10	–	–	10.7	7.5
	16	–	–	14	9.4
	20	–	–	–	10.5
	31.25	–	–	–	13.1
	62.5	–	–	–	18.4
	100	–	–	–	23.2
ACR	1	–	–	–	–
	4	–	–	–	40
	10	–	–	–	35
	16	–	–	–	30
	20	–	–	–	28
	31.25	–	–	–	23
	62.5	–	–	–	13
	100	–	–	–	4
Cable Length	Cable Type	Class A	Class B	Class C	Class D
	Category 3	2000	200	100	–
	Category 4	3000	260	150	–
	Category 5	3000	260	160	100
	STP	3000	400	250	150
DC Loop Resistance	–	560	–	40	40

Optical Fibers

Fiber-Optic Cable

	Multimode Fiber		Single-Mode Fiber	
	850 nm	**1300 nm**	**1310 nm**	**1550 nm**
Bandwidth (MHz-km)	200	500	–	–
Attenuation (dB/km)	3.5	1.0	1.0	1.0

Fiber-Optic Connecting Hardware

Connector Insertion Loss (dB)		0.5 average; 0.75 maximum
Splice Insertion Loss (dB)		0.3
Durability		500 mating cycles
Return Loss (dB)	Multimode	20
	Single Mode	26
Preferred Connector		SC

Fiber-Optic Link

Characteristic		Multimode Fiber		Single-Mode Fiber	
		850 nm	**1300 nm**	**1310 nm**	**1550 nm**
Bandwidth (MHz-km)		100	250	–	–
Attenuation (dB)	Horizontal (100 m)	2.5	2.3	–	–
	Building Backbone (500 m)	3.8	2.8	–	–
	Campus Backbone (1500 m)	7.4	4.4	–	–
Return Loss (dB)		20	20	26	26
Link Margin (dB)		11	11	11	11

A comparison of TIA and ISO link standards. Notice that the ISO standard is more relaxed than the TIA standard.

Link Level	Frequency (MHz)	NEXT (dB) min.		Attenuation (dB) max.	
		TIA	ISO	TIA	ISO
Category 5 Channel/Class D	1	60.0	54	2.5	2.5
	4	50.6	45	4.5	4.8
	10	44	39	7	7.5
	16	40.6	36	9.2	9.4
	20	39	34.5	10.3	10.5
	31.25	35.7	31.5	12.8	13.1
	62.5	30.6	27	18.5	18.4
	100	27.1	24	24	23.2
Category 3 Channel/Class C	1	39.1	39	4.2	3.7
	4	29.3	29	7.3	6.6
	10	22.7	23	11.5	10.75
	16	19.3	19	14.9	14

Appendix B
ABBREVIATIONS AND ACRONYMS

λ	wavelength
μm	micrometer
ACR	attenuation-to-crosstalk ratio
ASCII	American Standard Code for Information Interchange
ATM	asynchronous transfer mode
AUI	attachment unit interface
AWG	American wire gauge
BL	blue
BR	brown
c	speed of light (electromagnetic energy) in free space
CAP	carrierless amplitude and phase
CATV	community antenna television
CBX	computerized branch exchange
CD	compact disk
CRC	cyclic redundancy check
CSMA/CD	carrier sense, multiple access, with collision detection
DAS	dual attachment station
dB	decibel
dBm	decibel referenced to a milliwatt
EBCDIC	Extended Binary Coded Decimal Information Code
ECL	emitter-coupled logic
EIA	Electronics Industries Association
EMD	equilibrium mode distribution
EMI	electromagnetic interference
FCC	Federal Communications Commission
FDDI	fiber distributed data interface

FOTP	fiber-optic test procedure
FTP	foil twisted pair
G	green
Gbps	gigabit per second
GHz	gigahertz
IC	intermediate cross connect
IDC	insulation displacement contact/connector
IDF	intermediate distribution frame
IEC	International Electrotechnical Committee
ISO	International Standards Organization
kHz	kilohertz
km	kilometer
LAN	local area network
LCF-PMD	Low-Cost Fiber Physical Media Dependent
LED	light-emitting diode
Mbps	megabit per second
MC	main cross connect
MDF	main distribution frame
MHz	megahertz
MIC	medium interface connector
MLT-3	multilevel transmission (three level)
MMF-PMD	Multimode Fiber Physical Media Dependent
NA	numerical aperture
NEC	National Electrical Code
NEXT	near-end crosstalk
NIC	network interface card
nm	nanometer
NMS	network management system, network management station
NRZ	nonreturn to zero

NRZI	nonreturn to zero inverted
O	orange
OFSTP	optical fiber standard test procedure
OLTS	optical loss test set
OSI	Open Systems Interconnection
PABX	private automatic branch exchange
PBX	private branch exchange
PC	personal computer
PMD	Physical Media Dependent
RF	radio frequency
RJ	registered jack
RZ	return to zero
SAS	single attachment station
scTP	screened twisted pair
SMF-PMD	Single-Mode Fiber Physical Media Dependent
SNMP	Simple Network Management Protocol
SRL	structural return loss
STP	shielded twisted-pair cable
sUTP	screened unshielded twisted-pair cable
TC	telecommunications cross connect
TIA	Telecommunications Industry Association
TP-PMD	Twisted Pair Physical Media Dependent
TSB	Telecommunications Systems Bulletin
TTL	transistor-transistor logic
USOC	universal service order code
UTP	unshielded twisted-pair cable
V	volt
W	white

Appendix C
GLOSSARY

10BASE-2 Ethernet over thin coaxial cable at 10 Mbps.

10BASE-5 Ethernet over thick coaxial cable at 10 Mbps.

10BASE-F Ethernet over fiber-optic cable at 10 Mbps.

10BASE-T Ethernet over unshielded twisted-pair cable at 10 Mbps.

100BASE-FX Fast Ethernet over fiber-optic cable at 100 Mbps.

100BASE-T4 Fast Ethernet over Category 3 unshielded twisted-pair cable at 100 Mbps and using four cable pairs.

100BASE-TX Fast Ethernet over Category 5 unshielded twisted-pair cable at 100 Mbps and using two cable pairs.

100VG-AnyLAN A 100-Mbps network originally defined to use quartet signalling to operate over Category 3 (voice grade) UTP and offering a migration path for both Ethernet and Token Ring. Also supports Category 5 UTP, STP, and fiber.

access method The rules by which a network device gains the right to transmit a communication on the network. Common methods include carrier sense/multiple access, with collision detection; token passing; and demand priority.

asynchronous transfer mode A cell switching network using short, fixed-length cells operating at a basic speed of 155 Mbps, with other speeds including 25 and 51 Mbps to the desktop and 622 Mbps in the backbone.

attenuation A decrease in power from one point to another.

attenuation-to-crosstalk ratio (ACR) The difference between the power on a received pair to the crosstalk on the same pair, giving an indication of the margin between the noise level and signal level. The ACR typically decreases with increasing frequency.

backbone cabling Cable running between buildings or between telecommunications closets. Vertical cable. In star networks, the backbone cable interconnects hubs and similar devices, as opposed to cables running between hub and station. In bus networks, the bus cable.

balanced transmission A mode of signal transmission in which each conductor carries the signal of equal magnitude but opposite polarity. A 5-volt signal, for example, appears as +2.5 volts on one conductor and -2.5 volts on the other.

balun An impedance-matching device that also provides conversion between balanced and unbalanced modes of transmission. The term derives from balanced/unbalanced.

carrier sense/multiple access, with collision detection A networks access method used by Ethernet in which a station listens for traffic before transmitting. If two stations transmit simultaneously, a collision is detected and both stations wait a brief time before attempting to transmit again.

coaxial cable A cable in which the center, signal-carrying conductor is concentrically centered within an outer shield and separated from the conductor by a dielectric. The structure, from inside to outside, is center conductor, dielectric, shield, outer jacket.

cladding The outer concentric layer of an optical fiber, with a lower refractive index so that light is guided through the core by total internal reflection.

conduit A raceway of circular cross section. Also used more generally to indicate an enclosed cable pathway.

core The central light-carrying portion of an optical fiber.

crosstalk The unwanted transfer of energy from one circuit to another. Crosstalk can be measured at the same (near) end or far end with respect to the source of energy.

cutoff wavelength For a single-mode fiber, the wavelength above which the operation switches from multimode to single-mode propagation.

data connector A four-position connector for 150-ohm STP and used primarily in Token Ring networks.

DC loop resistance The total DC resistance of a cable. For a twisted-pair cable, it includes the round-trip resistance, down one wire of the pair and back up the other wire.

decibel The logarithmic ratio of two powers, voltages, or currents, often used to indicate either gain and loss in a circuit or to compare the power in two different circuits. $dB = 10 \log_{10} (P_1/P_2) = 20 \log 10 (V_1/V_2) = 20 \log_{10} (I_1/I_2)$. Power is most often used in premises cabling.

demand priority A network access method used by 100VG-AnyLAN. The hub arbitrates requests for network access received from stations, assigning access based on priority and traffic loads.

dispersion A general term for those phenomena that cause a broadening or spreading of light as it propagates through an optical fiber. The two main types are modal and material.

Ethernet A local area network using CSMA/CD for media access and operating originally at 10 Mbps and now at 100 Mbps over UTP, coaxial, and fiber-optic cable. IEEE 802.3.

far-end crosstalk (FEXT) Crosstalk that is measured on the quiet line at the opposite end as the source of energy on the active line. FEXT is not typically measured in premises cabling.

fiber distributed data interface (FDDI) A token-passing ring network originally designed for fiber optics and having a data transmission rate of 100 Mbps.

graded-index fiber An optical fiber whose core has a nonuniform index of refraction. The core is composed of concentric rings of glass whose refractive indices decrease from the center axis. The purpose is to reduce modal dispersion and thereby increase fiber bandwidth. The 62.5/125-μm fiber recommended for premises cabling has a graded index.

horizontal cabling That portion of the premises cabling that runs between the telecommunications outlet of the work area and the horizontal cross connect in the telecommunications closet.

horizontal cross connect The cross connect that connects horizontal cable to the backbone or equipment.

IEEE 802.3 The standards committee defining Ethernet networks; networks meeting these standards.

IEEE 802.5 The standards committee defining Token Ring networks; networks meeting these standards.

IEEE 802.12 The standards committee defining 100VG-AnyLAN; networks meeting these standards.

insertion loss The loss of power that results from inserting a component into a circuit. For example, a connector causes insertion loss across the interconnection (in comparison to a continuous cable with no interconnection).

intermediate cross connect A cross connect between the first and second levels of backbone cabling.

jumper A cable without connectors, typically used at the cross connect to join circuits. Similar to a patch cable (which has connectors).

laser A light source producing, through stimulated emissions, coherent, near-monochromatic light – an intense, narrow beam at a single frequency. Laser in fiber optics are solid-state semiconductors.

light-emitting diode A semiconductor that spontaneously emits light when current passes through it.

main cross connect A first-level cross connect for backbone cable, entrance cable, and equipment. The main cross connect is the top level of the premises cabling tree.

material dispersion Dispersion that results from each wavelength travelling at a different speed than other wavelengths through an optical fiber.

modal dispersion Dispersion that results from the different transit lengths of different propagating modes in a multimode optical fiber.

mode field diameter The diameter of optical energy in a single-mode fiber. Because the mode-field diameter is larger than the core diameter, it replaces core diameter as a practical parameter.

modular jack The equipment-mounted half of a modular interconnection.

modular plug The cable-mounted half of a modular interconnection. While available in 4, 6, and 8 positions, the 8-position version is specified for premises cabling.

multimode fiber An optical fiber that supports more than one propagating mode.

near-end crosstalk (NEXT) Crosstalk that is measured on the quiet line at the same end as the source of energy on the active line.

numerical aperture (NA) The "light-gathering ability" of an optical fiber, defining the maximum angle to the fiber axis at which light will be accepted and propagated.

Open Standard Interconnect A seven-layer ISO model for defining a communications network. Premises cabling deals with the lowest layer, the physical layer.

optical fiber A thin glass or plastic filament used that propagates light. The signal-carrying part of a fiber-optic cable.

optical time-domain reflectometry A method for evaluating optical fiber based on detecting and measuring backscattered (reflected) light. Used to measure fiber length and attenuation, evaluate splice and connector joints, locate faults, and certify cabling systems.

patch cable A cable terminated with connectors and used to perform cross connects. Similar to a jumper cable (which has no connectors).

patch cord A patch cable.

patch panel A passive device, typically a flat plate holding feed-through connectors, to allow circuit arrangements and rearrangements by simply plugging and unplugging patch cables. The feed-through connectors can have the same or different interface on either side.

plastic fiber An optical fiber made of plastic rather than glass.

plenum The air-handling space between walls, under structural floors, and above drop ceilings used to circulate and otherwise handle air in a building. Such spaces are considered plenums only if they are used for air handling.

plenum cable A flame- and smoke-retardant cable that can be run in plenums without being enclosed in a conduit.

quartet signalling The signalling method used by 100VG-AnyLAN, in which the 100-Mbps signal is divided into four 25-Mbps channels and then transmitted over different pairs of a cable. Category 3 cables transmit one channel on each of four pairs.

raceway Any channel designed to hold cables.

repeater A device that receives, amplifies (and sometimes reshapes), and retransmits a signal. It is used to boost signal levels and extend the distance a signal can be transmitted. It can physically extend the distance of a LAN or connect two LAN segments.

ring network A network topology in which terminals are connected in a point-to-point serial fashion in an unbroken circular configuration. Many logical rings are wired as a star for greater reliability.

riser A backbone cable running vertically between floors.

SC connector A fiber-optic connector having a 2.5-mm ferrule, push-pull latching mechanism, and the ability to be snapped together to form duplex and multifiber connectors. SC connectors are the preferred fiber-optic connector for premises cabling.

shielded twisted-pair (STP) cable A type of twisted-pair cable in which the pairs are enclosed in an outer braided shield, although individual pairs can be shielded. The most popular type of STP is IBM 150-ohm Type 1 cable.

single-mode fiber An optical fiber that supports only one mode of light propagation above the cutoff wavelength. Single-mode fibers have extremely high bandwidths and allow very long transmission distances.

Sonet Synchronous optical network, an international standard for fiber-optic-based digital telephony.

ST connector A popular fiber-optic connector having a 2.5-mm ferrule and a bayonet latching mechanism. ST is a trademark of AT&T.

star network A network in which all stations are connected through a single point. Star configurations tend to be reliable.

step-index fiber An optical fiber, either multimode or single mode, in which the core's refractive index is uniform throughout so that a sharp demarcation or step occurs at the core-to-cladding interface. Step-index multimode fibers typically have lower bandwidths than graded-index multimode fibers.

strength member That part of the fiber-optic cable that increases the cable's tensile strength and serves as a load-bearing component. Usually made of Kevlar aramid yarn, fiberglass filaments, or steel strands.

structural return loss A measure of the impedance uniformity of a cable. It measures energy reflected due to structural variations in the cable. A higher SRL number indicates better performance (more uniformity and lower reflections).

telecommunications closet An enclosed area housing telecommunications and network equipment, cable terminations, and cross connects. It contains the horizontal cross connect where the backbone cable cross-connects with horizontal cable.

telecommunications equipment room An enclosed area housing telecommunications and network equipment, distinguished from the telecommunications closet by its increased complexity and presence of active equipment.

thicknet A popular term for an IEEE 802.3 10BASE-5 network.

thinnet A popular term for an IEEE 802.3 10BASE-2 network.

time-domain reflectometry A method for evaluating cables based on detecting and measuring reflected energy from impedance mismatches. Used to measure cable length and attenuation, evaluate splice and connector joints, locate faults, and certify cabling systems. See *optical time-domain reflectometry*.

token passing A network access method in which a station must wait to receive a special token frame before transmitting.

token ring A token-passing ring network, such as 802.5 or FDDI. Often capitalized to refer specifically to IEEE 802.5 networks.

twisted-pair cable A cable having the conductors of individual pairs twisted around one another to increase noise immunity. Twisted-pair cables operate in the balanced mode.

velocity of propagation The speed of electromagnetic energy in a medium (including copper and optical cables) in relationship to its speed in free space. Usually given in terms of a percentage. Test devices use velocity of propagation to measure a signal's transit time and thereby calculate the cable's length.

wiring closet A general term for any room housing equipment or cross connects.

work area That area of the premises cabling where users are located; the area from the communications outlet to the equipment connected to the premises cabling. Loosely, an office, cubicle, and so forth.

Appendix D
CABLE COLOR CODES

CABLE COLOR CODES

4-Pair UTP

The preferred color code is for both backbone and horizontal four-pair cable. The alternative is for stranded patch cable only. The chart also shows the termination positions for modular plugs.

Pair	Preferred		Alternative (Stranded Patch Cable)		Connector Pins	
	Color	Abbreviation	Color	Abbreviation	T568A	T568B
Pair 1	White-Blue	W-BL	Green	G	5	5
	Blue	Blue	Red	R	4	4
Pair 2	White-Orange	W-O	Black	BK	3	1
	Orange	O	Yellow	Y	6	2
Pair 3	White-Green	W-G	Blue	BL	1	3
	Green	G	Orange	O	2	6
Pair 4	White-Brown	W-BR	Brown	BR	7	7
	Brown	BR	Slate	S	8	8

25-Pair UTP

TIA/EIA-568A color coding of 25-pair cable is according to ANSI/ICEA S-80-576, which is based on the industry-standard Western Electric scheme. Each group of five pairs has a sub-group color that is matched to a standard set of five colors. The table also shows the standard terminating 25-pair cable in miniature ribbon connectors and wiring blocks.

Subgroup Color	Pair	Color	Abbreviation	Connector Pins	
				Miniature Ribbon Connector	Wiring Block
White	Pair 1	White-Blue Stripe	W-BL	1	1
		Blue-White Stripe	BL-W	26	2
	Pair 2	White-Orange Stripe	W-O	2	3
		Orange-White Stripe	O-W	27	4
	Pair 3	White-Green Stripe	W-G	3	5
		Green-White Stripe	G-W	28	6
	Pair 4	White-Brown Stripe	W-BR	4	7
		Brown-White Stripe	BR-W	29	8
	Pair 5	White-Slate Stripe	W-S	5	9
		Slate-White Stripe	S-W	30	10
Red	Pair 6	Red-Blue Stripe	R-BL	6	11
		Blue-Red Stripe	BL-R	31	12
	Pair 7	Red-Orange Stripe	R-O	7	13
		Orange-Red Stripe	O-R	32	14
	Pair 8	Red-Green Stripe	R-G	8	15
		Green-Red Stripe	G-R	33	16
	Pair 9	Red-Brown Stripe	R-BR	9	17
		Brown-Red Stripe	BR-R	34	18
	Pair 10	Red-Slate Stripe	R-S	10	19
		Slate-Red Stripe	S-R	35	20
Black	Pair 11	Black-Blue Stripe	BK-BL	11	21
		Blue-Black Stripe	BL-BK	36	22
	Pair 12	Black-Orange Stripe	BK-O	12	23
		Orange-Black Stripe	O-BK	37	24
	Pair 13	Black-Green Stripe	BK-G	13	25
		Green-Black Stripe	G-BK	38	26
	Pair 14	Black-Brown Stripe	BK-BR	14	27
		Brown-Black Stripe	BR-BK	39	28

25-Pair UTP *(continued)*

Subgroup Color	Pair	Color	Abbreviation	Connector Pins	
				Miniature Ribbon Connector	Wiring Block
Yellow	Pair 16	Yellow-Blue Stripe	Y-BL	16	31
		Blue-Yellow Stripe	BL-Y	41	32
	Pair 17	Yellow-Orange Stripe	Y-O	17	33
		Orange-Yellow Stripe	O-Y	42	34
	Pair 18	Yellow-Green Stripe	Y-G	18	35
		Green-Yellow Stripe	G-Y	43	36
	Pair 19	Yellow-Brown Stripe	Y-BR	19	37
		Brown-Yellow Stripe	BR-Y	44	38
	Pair 20	Yellow-Slate Stripe	Y-S	20	39
		Slate-Yellow Stripe	S-Y	45	40
Violet	Pair 21	Violet-Blue Stripe	V-BL	21	41
		Blue-Violet Stripe	BL-V	46	42
	Pair 22	Violet-Orange Stripe	V-O	22	43
		Orange-Violet Stripe	O-V	47	44
	Pair 23	Violet-Green Stripe	V-G	23	45
		Green-Violet Stripe	G-V	48	46
	Pair 24	Violet-Brown Stripe	V-BR	24	47
		Brown-Violet Stripe	BR-V	49	48
	Pair 25	Violet-Slate Stripe	V-S	25	49
		Slate-Violet Stripe	S-V	50	50

Multifiber Cables

TIA/EIA-598A defines color coding for multifiber cables. Either the fiber or the cable tube can be colored.

Fiber	Fiber/Tube Color	Fiber	Fiber/Tube Color
1	Blue	13	Blue-Black Stripe
2	Orange	14	Orange-Black Stripe
3	Green	15	Green-Black Stripe
4	Brown	16	Brown-Black Stripe
5	Slate	17	Slate-Black Stripe
6	White	18	White-Black Stripe
7	Red	19	Red-Black Stripe
8	Black	20	Black-Black Stripe
9	Yellow	21	Yellow-Black Stripe
10	Violet	22	Violet-Black Stripe
11	Rose	23	Rose-Black Stripe
12	Aqua	24	Aqua-Black Stripe

STP

The following is the color code information for 150-ohm shielded twisted-pair cable.

Pair	Color Code
Pair 1	Red
	Green
Pair 2	Orange
	Black

Appendix E
FURTHER READING

What follows is a brief list of books that expand on many of the issues covered in this book. The list leans toward application-oriented material.

Equally important are the standards that drive premises cabling. Standards can be ordered from:

American National Standards Institute (ANSI)
430 Broadway
New York, NY 10018
(212) 642-4900

Global Engineering
1990 M Street, N.W.
Washington, DC 20036
(800) 854-7179
(202) 429-2860

Standards

TIA/EIA-606: Administration Standard for the Telecommunications Infrastructure of Commercial Buildings.

ANSI.EIA/TIA-568A: Commercial Building Telecommunications Cabling Standard.

EIA/TIA-569: Commercial Building Standard for Telecommunications Pathways and Spaces.

TIA/EIA TSB-67: Transmission Performance Specifications for Field Testing Twisted-Pair Cabling Systems.

ISO/IEC 11801: Generic Cabling for Customer Premises.

Books

Sergio Benedetto, Ezio Biglieri, and Valentino Castellani, *Digital Transmission Theory* (Englewood Cliffs, N.J: Prentice-Hall, 1987).

BICSI, *Telecommunications Distribution Methods Manual* (Lexington, KY: Building Industries Consulting Services International, 1994).

Frank J. Derfler, Jr., *Guide to Connectivity*, 3rd ed. (Emeryville, CA: Ziff-Davis, 1994).

James Martin, Kathleen Kavanaugh, and Joe Leben, *Local Area Networks: Architectures and Implementations*, 2nd ed. (Englewood Cliffs, NJ: PTR Prentice-Hall, 1994).

William Stallings, *Data and Computer Communications*, 4th ed. (New York: Maxwell Macmillan International, 1994).

Donald J. Sterling, Jr., *Technician's Guide to Fiber Optics*, 2nd ed. (Albany: Delmar Publishers, 1993).

Index

V